THE SECRET LIFE OF SCIENCE

THE
SECRET
LIFE OF
SCIENCE

How It Really Works and Why It Matters

JEREMY J. BAUMBERG

PRINCETON UNIVERSITY PRESS
Princeton and Oxford

Copyright © 2018 by Princeton University Press

Published by Princeton University Press,
41 William Street, Princeton, New Jersey 08540

In the United Kingdom: Princeton University Press,
6 Oxford Street, Woodstock, Oxfordshire OX20 1TR

press.princeton.edu

Jacket design by Jason Alejandro

Library of Congress Cataloging-in-Publication Data

Names: Baumberg, Jeremy J., author.
Title: The secret life of science : how it really works and why it matters /
Jeremy J. Baumberg.
Description: Princeton : Princeton University Press, [2018] | Includes index.
Identifiers: LCCN 2017039857 | ISBN 9780691174358 (hardback : alk. paper)
Subjects: LCSH: Science—Social aspects. | Science—Methodology. | Scientists—
Training of. | Communication in science. | Discoveries in science.
Classification: LCC Q175.5 .B395 2018 | DDC 306.4/5—dc23
LC record available at https://lccn.loc.gov/2017039857

British Library Cataloging-in-Publication Data is available

Printed on acid-free paper. ∞

Printed in the United States of America

1 3 5 7 9 10 8 6 4 2

To my wife and daughters,
for combining love with the science.

CONTENTS

ACKNOWLEDGMENTS

I am by most measures a very well funded and successful scientist, becoming a professor at thirty and a fellow of the Royal Society before I was forty-five. But I am not a social scientist, so this book is an aberration in my life. It emerged initially from a sabbatical that I took in San Sebastian in the Basque Country of northern Spain in 2012. Every morning I ignored all my science research and administration tasks, as well as my hosts, and sat down asking myself questions about the system I find myself in, and looking for the answers. My career has been relatively unusual in science, mixing experiences in different cultures (the United States, Japan, Spain, the UK) with significant time in industry (IBM, Hitachi, and several spin-off companies). This is perhaps what has prompted me to question the makeup of the system I find myself in. But many of the questions I was asking did not have satisfying answers, and particularly, why the scientists I come into daily contact with do things in certain ways. Part of this book is an attempt to voice this. As you will see it is also partnered by a blog that aims to continue the exploration, at www .sciencemonster.com, to which I hope you will contribute.

The other motivation for writing this book has been my disquiet over many years about the typical view most people have about how science works, and what scientists do. This is only natural because they are fed this view in everything they typically read about science and technology. Even scientists in my experience rarely share the hidden side of their lives, this slight craziness about how it all works. There are many vested interests in the science ecosystem, and it helps to understand how these all work for and against each other.

However this book is in no way antiscience. In many countries some people call into question the very purpose and method of science (such as in the United States), to avoid it gaining an authority that takes away from others claims to power. I am convinced that the scientific process is extremely robust, a fantastic way to invest for the future of our societies, and an ultimate good. Of course, that is why I am a scientist. But this is not for me a sufficient reason to shy away from questioning how it works and how it develops. I am hoping these discussions will interest anyone who has the faintest collision with science, whether as a reader of science, a writer involving science, a funder (all of us), a civil servant, a politician, or even those questioning the role of science in our lives. All the scientists who I have discussed my ideas for this book with over the last few years have unfailingly asked to read the results of my investigation because it makes them realize how little they know of their own system—I hope that they won't be disappointed.

Many people have supported me along the path. The University of Cambridge, my department, the Donastia International Physics Center and the Spanish National Research Council (CSIC) in San Sebastian, and Jesus College Cambridge have all provided resources of one sort or another, including the rare gift of tranquil thinking spaces. My research group, many young researchers and administrators across the university, and colleagues have engaged enthusiastically and made suggestions. In particular a number of readers have given me excellent feedback, including specifically Christophe Febvre, Rob Findlay, Charlie Constable, Simon Schaffer, and Richard Jones, along with many others who devoted their precious time and attention to discussions. I also am indebted to the valuable support from Ingrid Gnerlich and many others at Princeton University Press. I should note that ideas exposed here are not others' responsibility—they result from my personal views alone. Finally my wife and children have been unstinting in encouraging me and allowing me to devote time away from them. This book is dedicated to them for making my science livelihood a full life.

THE SECRET LIFE OF SCIENCE

1

QUESTING SCIENCE

What science gets done?

Why do we do science? As a teenager, I took the answer for granted. We explore, as we have always explored, our own world's peaks and rifts, forests and plains, as we clamor to explore other planets and comets. Now that I deal in science I know this cannot be enough, cannot tell us why we all are forced to learn science in school and insist that our children do too, cannot tell us why we spend so much of our financial resources and invest our hopes in science. With explorers there were always only a few, and yet with scientists we have now grown a vast abundance.

"Why," my daughter asks at breakfast this morning, "why is it bad to drain marshes if it stops mosquitoes breeding?" Nowadays a quick web search with her rapidly shows how much science has accumulated behind this query. We read online studies that summarize why drainage can be both good and bad, short term and long term, why every component in the natural cycle is linked into so many others that simple predictions don't often work out. For every "why?" there are lines of Russian dolls stretching out to the horizon, waiting to be opened up to reveal new chambers of information, each gleaned over decades of careful scientific investigation.

We take our science for granted . . . that scientists are everywhere around us, that their breed isn't dying out or losing its way, or that it isn't driving its convoys up blind alleys. We assume science will keep on trotting out its miracles that get ever further from our human experience, and deeper into imaginative territories. But how do we know it is in good shape? You pay for it in multiple ways, but do you care for it, do you nurture science? What science are you being offered, which fragments from the vast terrain?

Our societies have discovered that science is a powerful stimulus. It shocks and challenges us, it creates visionary ambitions, and its dopamine-fuelled lure is a natural focus for the human mind that loves "aha!" We trust scientific results that stand the brutal tests of criticism and time. We are so confident in this that we believe any self-aware civilization, anywhere in the universe, would agree with our own established scientific principles. On the other hand, the *way* that this self-aware civilization went about studying science would likely be completely baffling to us. The scientific *system* we have created is not a foregone conclusion. Ancient cultures did not have formal organizations to explore the natural world—instead it was left to the curious few who carved themselves temporarily safe niches within their societies. The construction of formalized institutions such as universities, around the world, which foster learning and organized discussion, is one of the many unremarked miracles of our current lives.

Just *what* science gets done is much more determined by our culture than is the objective truth of our science. The essential need for repeatability and testability in science means that any crack or flaw in the network of interconnected ideas is mostly quickly identified, but *what* we decide to expend our limited resources on within this infinite structure is not obviously clear.

Approaching the scientific enterprise from the outside, you might imagine that what gets done is formed on agreement between leaders, or emerges from a sort of signposting by nature that is apparent to anyone looking dispassionately at each scientific subfield. My own discussions with public focus groups often bring out their belief that "surely *someone* is directing the entire science operation." But the truth is much more complicated, and far closer to the way we steer our economy, buffeted by inevitable fashions, gurus, and cycles of booms and busts caused by our social interactions. It is virtually impossible in any country to identify *who* made a decision to do *this* piece of science. On the other hand, most countries are effectively locked into competition to do the "best" science, by trying to identify it and fund it.

As I tried to answer my own questions about the web in which I am involved, I became ever more fascinated with accounting for why I do what I do, and surprised by what this investigation says might be usefully changed. This book tries to capture what influences the science that gets done, and why it matters. To do that we have to look at the motivations of everyone involved, beyond those of just the researchers themselves. The view I will elaborate is that it helps to consider the full *ecosystem of science*. Ecosystems help us see how chunks of the natural world balance themselves, through relationships between many natural organisms all at once. The different parts of the science system act similarly as interrelating organisms. Intense competition—not just between scientists, but between research journals that publish results, universities that house researchers, newspapers, governments, subject areas, and many other contestants—favors survival skills. The resulting ecosystem of science has its own personality and temperament, which leads it toward a future that we might all want to have our own say in.

This book maps my own explorations, asking questions that both scientists and nonscientists have been confused by, and I am startled to find are not well known. Although it has been useful to be a practicing scientist, this is not necessary to appreciate the stories that emerge. In their telling, we might actually understand what it is that our societies want scientists to do (figure 1.1).

In starting to map out the territories of this book I realized that there really is no good place to find a description of the way science actually works. This landscape turns out to be complicated, interesting, and connected. And yet it should also influence how we read explanations of scientific results, visions, funding, and ambition. Many of the issues I raise can be found in fragments, across blogs and letters pages, and discussed by scientists, sometimes widely. However each minidebate is part of a wider frustration that I sense in scientists

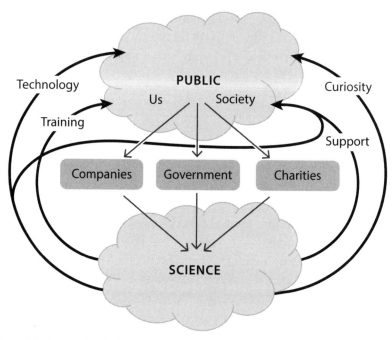

Figure 1.1: The bargain of science.

themselves, of being trapped in a system not of their making. And even worse, they feel that they are powerless to modify this system, because its meshing cogs are so finely aligned that no change is independent of every other part. This book is thus also a stimulus to restore balance: for nonscientists to regain some insight and control into what happens with their funds, and for scientists to regain some insight and control into how it happens.

WHY SHOULD WE CARE?

We pay for science. We accept the tantalizing gifts it provides, though we are sometimes nervous about their effects or about their ethics. And science will frame our future, the future for our children, and the long, deep future for humans on this planet. On average we each pay only a few hundred dollars a year to support science, not very much compared to what we burn up just driving around. But this is one of the best investments we make in our collective future, returning far more than any other opportunity for our savings. How this investment is spent on our behalf is a very peculiar thing, and this is another story of this book. The increasing wealth of our societies has nurtured the growth of a new ecosystem, both insulated from and buffeted by wider forces across society, in which science is mined, amassed, archived, and expanded.

The book is planned as a survey, focusing on different parts of the system in turn to see how the overlapping spheres of influence mesh together to create a science ecosystem. I start with the science itself and show how scientists are fragmented between a pair of interdependent roles. I also ask how realistically we can view science as its own ecosystem, and how this raises questions to consider about science. To ground our discussions I review motivations for doing science in chapter 3, using the lens of what scientists have thought most important over the past half century. In mapping the chains of influence, the next three chapters consider in turn how scientists present their newfound knowledge, how they hear about it from each other, and finally how we all across society get to hear about it through the media. These illuminate the flows that bind the spheres of knowledge, people, and media around scientists themselves. Later in chapter 7 I explore what resources go into science and who controls where this money flows. How the people who make this all happen are shaped along the way is sketched out in chapter 8 to complete the survey of the overlapping spheres and their myriad internal competitions. In chapter 9, I pull all this together to consider how evolution in this landscape is responsible for its current situation, before making some suggestions about ways we might change it in the final chapter, offering tentative prescriptions. Rather than a polemical stance, my main aim is to describe what the state of science really is and how it works.

I am a practicing scientist, and thus deeply implicated in the system, but I had an early career in industry and so already approached research with one eye refracting the perspective of an outsider's prism. My aim is to write as an outsider might look in, as a science sociologist of the mainstream, laying bare the construction of this science system. Popular accounts of science instead tend to focus on larger-than-life personalities, early historical perspectives, and to lay out compilations of mythical and anecdotal stories to provide color and moral fables. I am going in a different direction, both because my theme is about the wider interactions, but also because I want to capture what we have now and where it might be going.

Along the way I find many questions whose answers I felt I ought to know. How many scientists are there in the world? How many physicists, chemists, biologists, or engineers? How many more or less will there be in ten years? Of the thousands of papers published each day, why do a few get into our newspapers and Twitter streams? Who chooses for us, and why? Who chooses what sort of science gets done? What do scientists read? How do they choose what science to do? Are there too many biologists in the world, or too few? What sets the size of a science conference? These are not part of my background knowledge, nor that of most scientists.

Science is in rude health, never having been better funded or producing more results. Any system has its flaws though, and in science many are well known: influences of money and self-interest, sporadic scandals of distorted or false data and conclusions, sensationalisms of wild hype or wild personalities, distorted reporting from the media circus, or individuals stealing credit. All these are persistently aggravating to the cause of science, but they do not fundamentally undermine it because of the self-correcting way that it works.

However another set of deeper worries will emerge from my survey here, ones that are an implicit part of the system of science itself. The contested interactions that are integral to the science ecosystem are inflating competition between its different components. This ever tightening set of global competitions skews how the system works and evolves. Through this book, I hope to show why I am no longer quite so convinced about the globalization of science and why I believe science might be expanding too much. These competitions have other effects that will become apparent, such as reducing the diversity in how science is funded and appreciated. All these pressures focus on the scientist, who, contrary to conceptions of airy independence from societal pressures, is at the nexus of some of the most intense global struggles. But to start at the beginning, we first need to ask what we mean by science.

2

WHAT IS SCIENCE?

Most people have a comfortable familiarity with the idea of science. My first task though is to prise apart this solid-seeming notion of what it is, and to locate two contrasting types of science inquiry and what drives each one. To understand the ecosystem as a whole, I then want to tackle how many researchers there are at the moment, and how this is changing. This tribe of scientists is the core of the science ecosystem, and I will discuss how they divide themselves within it. Finally I will discuss how the science ecosystem might work, mapping out the different parts we will visit on this tour.

THE ROLES OF SCIENTISTS

SCIENCE IN OUR SURROUNDINGS

Look around you now. What do you see that a spirited woman teleported forward through time from prehistory might recognize or understand? Tables, floors, sockets, screens, lights, windows. All are absent from the natural world; all offer intricate wonders. What would intrigue your prehistoric visitor most? Each object has myriad stories to tell, ways of understanding culture, history, science, and technology, all blended.

Correspondingly, how do your surroundings intrigue *you*, and what questions emerge from where your gaze alights? How is that glittering shard of light hitched by wires to the wall spilling colors over the room? Perhaps you are puzzled by the inaudibly chattering microwaves from your mobile devices ricocheting off your friends and your floors, passing tenuously through walls. How come this metal spoon feels hot, while the porous ceramic of your coffee mug that holds it is not. How are the clumped words on this page leaving indelible chemical marks in your brain, for you to reconsider in your pauses?

Science is not only so concretely apparent. Perhaps you are more concerned about invisible fears raised by hectoring all around us. What is really aloft in the air we are breathing, visible as that sulfur yellow-white horizon smog. How much guilt did that ear-tautening plane journey add to how you feel about our use of planetary resources. Is an article you read really showing a promising cure for cancer?

These ways of viewing the world, asking the "why is it," "why does it do this" questions, are the natural pesterings of a scientific stance. Mostly our questioning attitude starts near home. Why we are ticklish, why only some of us, and then what tickling might be for. How do dogs navigate over long distances so well? Such questions often still don't have answers that fully satisfy us. But science questions roam much further. While existing natural things just are what they are, interlocking into an accepted part of our worlds, new invented devices are weirdly fascinating. Thumbnails that store all our music. Glittering black solar panels that feed our energy habit. The dazzle of technologies that make miracles, fleeting glimpses of their inner science only hinting at the underpinnings that drive them.

How do these ways of seeing our surroundings relate to "science"? When are we actually doing science? And what does it mean to be doing science? The pages of newspapers and websites offer a glimpse of the problem of defining science and categorizing its activities. As I write this, my web feed has pieces about the particle accelerator at CERN under "science," polar bear feeding ranges are "climate change," and discussions about battery storage concepts are "technology." Are these all science too? Can "scientists" work on technology, or the environment? To deal with this fuzziness I will first revisit where science emerged from.

WHERE ARE OUR STEREOTYPES?

Our idea of what science encapsulates constantly changes. From the sixteenth century, "natural philosophy" labeled those eccentrics musing about the physical realms and its increasingly fertile experimental methods. While the label *science* came from the Greek for the *whole* of knowledge, it was soon captured by those focusing solely on the natural and physical domains. By the mid-1800s "scientists" were seen as those who sought knowledge from nature only. A hundred years ago, we had emerging stereotypes: lone scientists devoted to understanding an enigmatic Nature, gluttonous collectors labeling abstruse specimens from one of Nature's rich niches, entrepreneurial technologists hammering out fixes to problems that people will pay for. But things have moved on a century, and while the appetite for peeking over the shoulders of scientists has become even more compelling, to funders and the public alike, our old categories divide poorly.

Scientists and technologists constantly influence each other. Greek-speaking scholars fleeing the Turks influenced an Italian humanist movement keen to rediscover classical literature and revitalized science knowledge (idolized now as the "Renaissance"). When the Black Death made labor expensive in the fourteenth century, technology grew in importance, milking this nascent science for manufacturing materials and devices that could turn a profit. Gutenberg's printing advances in the fifteenth century distributed this knowledge widely, while the improving ship technologies enabled the collection of new discoveries from far and wide. Systemization of this arrayed knowledge cried out for the explanatory power of science. Industrialization accelerated these cycles of science and technology, with huge effects in alleviating the drudgery of living by the twentieth century.

Our current crop of scientists inherits these multiple motivations. Typically, scientists are divided into "pure" and "applied," taken to mean those who study the natural world, and those who try to create useful things from this knowledge. Note firstly how emotive these labels are since "purity" emphasizes a greater virtue, in opposition to polluting influences. In society it is associated with ideas such as racial or sexual purity, defining implicitly an undesirable dark side. To understand what motivates scientists, I believe that it is more useful to employ a different (and more neutral) dichotomy, which emphasizes the roles they take when doing science. I will describe these as "simplifiers" and "constructors," and show how they both create fundamental scientific knowledge.

INTRODUCING SIMPLIFIERS

Consider the bands of physicists who survey the nature of different subatomic particles, and what component particles these in turn might be made up of. Some ask "why do the particles have this mass?" or "why do they collide in this manner to produce a shower of new particles?" They strip down complexity to identify more basic component properties, ways that allow us to explain what we see in simple concepts. Such scientists I will call *simplifiers*, though related epithets might be "explorers," "delvers," or "deconstructors." Their relation to the world is of opening the box of magic and trying to understand how it works, and they are the reductionists of science.

It is not obvious at the outset of any scientific journey that a simplifiers' approach might work at all. Simplifications seem to be possible only because the underlying physical behavior of the world is really governed by laws that conserve various underlying essences. We know that energy is conserved, momentum is conserved, and the number of atoms is conserved. These laws can all be broken in various ways: we can convert energy to mass and back; we can lose atoms from a body by abrasion or chemical conversion into gases; we can convert atoms from one to another in nuclear reactions. What is most useful is that the underlying reasoning behind these laws allows us to convert questions we have now into close correspondence with questions we have asked before, and this allows us to form a scientific intuition. Each situation is not new, but related to those considered before. Without such order, intuition is very difficult to achieve, swamped by a morass of details and information fragments. Simplifiers are at work within every domain of science.

DO SIMPLIFIERS KILL BEAUTY?

Simplifiers are associated with several different emotional reactions to science. They have long been a target for romantics of the world, who see the uncovering of inner workings as inhumanly mechanistic and killing the essence of outward beauty. To some, knowing the simpler reasons that animate complex phenomena (such as why rainbows are colored) is seen as stripping them of power, life, and energy. Similar allegations are made about the destruction wrought by analyzing poetry, literature, or art.

On the other hand, simplifiers are seen as great explorers who find new unknowns in the universe to boldly go beyond. This pioneering aspect particularly unfolds in large-scale megascience projects such as particle accelerators, where the stress is on the romance of breaking boundaries of knowledge and providing new understandings of our human place in the cosmos. They are the knights who vanquish the implacable dragons, overturning past theories. This long-standing fight-back by simplifiers inverts the charge of killing the world when pinning its beauty to a board, like lifeless butterflies. Every person stares at stars and can be awed. Knowing that huge balls of squeezed fusing gases glare out at us blowing off immense canopies of glowing streamers only adds to our awe (figure 2.1). While reduction is indeed part of the task of simplification in science, it is not a reduction in content but a transformation in the task of comprehending, from voluminous fact to stacking and interleaving of scientific principles and implications. This is what is meant by understanding some piece of science.

COMPRESSING INFORMATION

For simplifiers, a satisfactory outcome in a scientific field is "information compression." They aim for fewer facts and more connecting description (often in the form of equations defining relationships) to directly account for the world. Sometimes they can show why information compression is not possible, such as in predicting weather, which we are used to knowing only days in advance. In other areas such as predicting the large fluctuations in the stock market, we find it hard to accept such unpredictability. It is interesting that situations

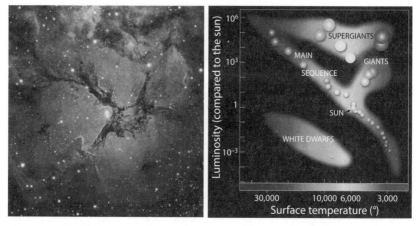

Figure 2.1: Ways of inspiring. *Left*: The ruby-hued Trifid nebula several light years in diameter. The hydrogen gas core is heated by hundreds of brilliant young stars causing it to emit red light (just as hot neon gas glows red-orange in illuminated signs). Image credit: R. Jay GaBany, Cosmotography.com. *Right*: The evolution of stars can be tracked as they cool (from blue-hot on the left, to red-hot on the right) and fade in luminosity. Image credit: © 2017 ESO.

where simplifiers currently hold most authority are now those parts of science *least* amenable to repeated human experimentation, and those that possess a very large number of complex components. These include cosmology (why the distribution of stars, galaxies, quasars, and black holes appears as it does), biology (what is alive, how do cells work, how does the brain work, and how do we experience consciousness?), and particle physics (why is there so little antimatter in the universe, what gives the most basic particles known their mass and charge, and what do we really mean by mass and charge?). Challenges exist for every discipline in situations where full control remains difficult, and observations are problematic. Simplifier science is easiest and has worked best when we can twiddle all the knobs on the box, and if we break it we can try another box. Where this has been possible, simplifiers have already swept through triumphant.

Science can get into difficulties though when testability or repeatability is challenging. If scientists have a theory about how the world works, they need it to make predictions. But if such predictions play out over millennia and involve scales beyond our control (such as collisions between stars) then scientists cannot set up an experiment and wait to check that their expectations are fulfilled. Nor can they try this experiment several times to check it works, even with small variations in the initial setup conditions (to test if the predictions are *robust*). Similarly it is not yet possible to set up a particular set of thoughts or new memories in someone's brain and observe the repeatable electrical firing of billions of neurons. Problems with testability and repeatability are found in disciplines and questions that are precisely the ones which have *not* proceeded so far, because we cannot travel experimentally in lockstep with our theories. Instead scientists have to resort to observations in which they catch specific occurrences in the act (looking everywhere in the cosmos with telescopes to find a few galactic collisions), or they wait until something they want to test comes along (watching a brain seizure develop). Such a strategy depends on the richness and density of components under study (stars, neurons). This makes the strategy rather more difficult for improbable events or those in sparse situations, such as abrupt climate changes, stock market crashes, supernovas, or the development of life on a planet.

As a result we have a distinct class of hard science problems in which simplifiers are currently richly represented, but that are hardly amenable to testability. These must be set against a huge number of scientific areas in which the simplifier approach has already been enormously successful (such as electronics, or molecular bonding). The result of those has been a highly compressed information web that is the seed for much new science. Most of these mapped areas are ones with multiple types of amenable experiments, giving great assurance to the credibility of our scientific understanding. They allow confident predictions about the world, which will be important for the other way of doing science discussed below. It is the success of simplifiers in compressing information that leads to the second role for scientists that I now describe.

INTRODUCING CONSTRUCTORS

Once simplifiers can confidently predict how a chunk of the world works, a new phalanx of questions emerge that head in completely different directions. "How can we use this breakthrough, not just to simplify, but to build something?" Descriptions based not on assemblies of facts, but on toolboxes harnessing interactions and conservation laws, inherently give birth to a different sort of creativity that generalizes these to new arenas. "What happens if I do this to that?" or "how might this operate if I heat it?" Notice that this sort of science is not directly asking how an aspect of nature works. It is already making constructs, mostly never found before in nature, in which questions of "how it works" are always twinned with "how can we make it work differently?" For this reason, I will refer to such scientists as **constructors**, though related descriptions might be "builders," "composers," "creators," "makers," or "assemblers." They may not be applying this knowledge to any concrete goal (so "applied" science is not a good description), and they may be theorists or experimentalists.

Constructor science is based on initial simplifier discoveries and would simply not be possible unless those discoveries were robust in almost every way. Imagine the scientific landscape as a city of buildings. The architectural toolbox produced by the simplifiers includes cables, concretes, columns, struts, and beams. Constructors put these together in imaginative ways testing the principles of rigidity, stress, and design rules in arches, domes, and buttresses to their limits, as well as sculpting elegant forms. Any flaw in understanding the components and how they might go together is rapidly uncovered by the resounding sound of buildings collapsing. The reason we don't hear a cacophony of scientific rubble crashing to the ground is the sheer robustness of most of our knowledge. One might interpret this peaceful city as evidence of resistance to new forms, conservatism, or sloth. But actually the serene silence pays tribute to the success of science. Computers rely on our understanding of complex physics being correct and remaining correct, trillions of times every second, everywhere.

This web of knowledge that is the historical accumulation of all the science-building work of the past is our legacy to future societies. It is composed of all the written-up reports of scientists published in journals, their collected technologies and equipment, the know-how of all scientists alive, and the network of relationships between all these. The web is built both by the simplifiers and the constructors, who interact with each other to stimulate new science, and who

SIMPLIFIERS **CONSTRUCTORS**

Figure 2.2: Contrasts between *simplifiers* and *constructors*.

together expose tears in the fabric of knowledge that need attentive mending. Because the web of knowledge is dense and taut in some areas of science, it is very difficult to disturb radically, and there are few spaces still hidden from view. On the other hand, knowing this web in some detail still does not make science into a programmatic activity that can be continued by rote. Instead constructor science produces an astonishing litany of emerging surprises, which is perhaps the second miracle of science.

These two types of science roles, simplifier and constructor, are both needed for creating the ecosystem of science (figure 2.2). But why should it be that understanding the components (in simplifier fashion) does not open up the complete architecture of all science to us? The reason is that perfect comprehension at one level of abstraction does not give enough understanding to help know what might happen when we combine components at this first level into a more complicated assembly. The rigidity of bricks does not make arches an inevitability. This is the concept of *emergence*, what the physicist Philip Anderson has sloganed "More, is Different." It blows apart the idea that knowledge of the subatomic construction of quarks can tell us much about the biochemistry of life. The layering of the different scientific disciplines is based on this principle. We might have predicted self-replicating molecules long ago, but it was not *inevitable* from knowing chemistry. We stumbled across it. Constructors are those scientists who drive this emergent science.

Constructors are very common in science and have always been so historically. Guglielmo Marconi, who was the first to transmit messages by radio waves from the UK to the United States, followed on from a generation of simplifier work by scientists such as James Clerk Maxwell, who uncovered the basic equations for electromagnetic waves, which include light, X-rays, and microwaves, as well as radio waves. While sometimes the same scientist takes both simplifier and constructor roles, these different drives are often separated between people.

FERTILE MINIWORLDS

The handover from simplifier to constructor science in reality is very subtle. Often, in order to test whether a simplifying scientific idea is correct, an austere model is proposed, and this in turn suggests a new series of experiments on a stripped-down set of small components. To understand a poison, a particular chemical component might be injected into a cell, or a specific organ, which is then observed closely. To understand the earth's inner magnetic field, we subject tiny diamond vials of iron to enormous pressures and temperatures. Such experiments are very important in revealing a phenomenon in its simplest setting, abstracted as much as possible away from messy context, with the minimum of complicating distractions, and they are the essence of the simplifier's armory. However the very act of removing a phenomenon from its original setting, and corralling it within a protected nest in which it is coaxed to operate as close as possible to its model, inherently creates new possibilities and constructs new science.

A traditional realm for this use of model systems is the "Gedanken" or thought experiment, actually a theoretical approach coined by Hans Ørsted, and dramatically exploited by Erwin Schrödinger and other pioneers in developing their quantum theories a century ago. Concerned with motion at the most microscopic level, quantum mechanics was initially used to explain the specific palette of colors emitted by aroused atoms. As well as its success in accounting for the observed polychromatic glows (for instance from sodium street lights), these theories suggested that just watching a particular particle would actually change how it moves. The impossibility of accomplishing such experiments in the 1920s focused attention on an analytical approach that constructs "thought experiments" and then considers their implications. These completely idealized experiments nevertheless suggested such bizarre and wondrous conclusions that they stimulated a raft of new theories. Eventually improved technologies enabled real experiments, which then agreed with the predictions almost entirely. Now ever more beautiful realizations of these Gedanken experiments help young scientists (and us all) to be convinced of the essential truths of the quantum mechanical theories. We are forced to see splattered onto a detector the waves from chunky protein molecules that each somehow squeeze through two close-spaced pinholes at the same time. Even such big objects defiantly have a wave-like aspect.

While simplifiers verify the truth of ever more sophisticated tests of quantum mechanics on individually observed particles, building such ideal constructs has stimulated constructors to envisage ways to exploit quantum systems. An entire subfield has emerged around the use of quantum mechanics to encode information that bucks the pervasive digital paradigm of binary zeros and ones (figure 2.3). Quantum information is recorded as *simultaneous* combinations of one and zero, as if future electronic circuits could exist switched into both "on" and "off" states at the same time, calculating both situations at once (and thus faster). Related research measures single

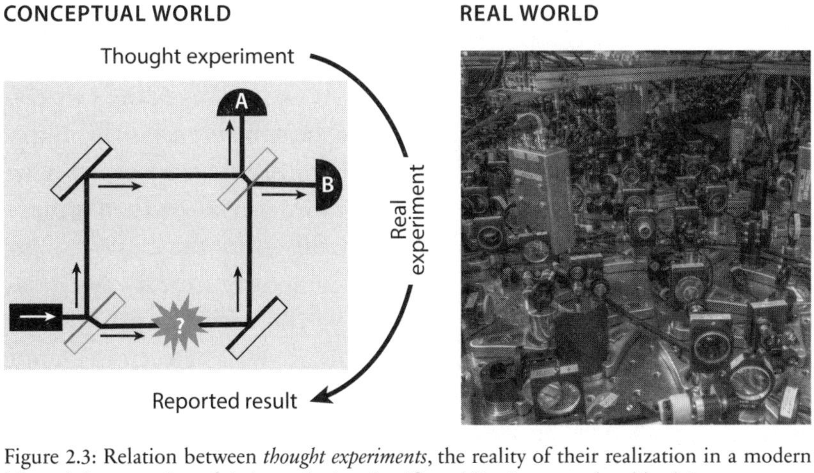

Figure 2.3: Relation between *thought experiments*, the reality of their realization in a modern lab, and the reporting of their results in scientific publications translated back into a conceptual framework.

electrons, or discusses ultrasecure quantum cryptography, ensuring no one can read and duplicate critical information you are sending (such as your bank password), or explores the teleportation of images and even possibly objects. Quantum instruments are now built that measure the gravitational pull of underground oil reserves with exquisite sensitivity, and can also detect the local electrical currents directly in your sparking brain. Thus are the models of simplifiers now reborn as tools for constructors.

WHAT DO CONSTRUCTORS DO?

This gives a feeling for how constructors work. They are more interested in building a new model system and seeing new behaviors emerge than in answering questions about a system provided by nature. Sometimes they have an idea, but in working through their hunch they stumble across something completely unexpected. This is why they find it harder than simplifiers to explain why they do what they do, since they are not "curing cancer" or addressing similar-scale grand challenges. In fact they often produce the critical breakthroughs that are most helpful in tackling such grand-challenge problems. They cannot say beforehand that this is likely to happen, but only what they hold in their sights and its possibilities. A good example is the material science community, who are spread widely across companies and universities, with most practitioners trying to achieve diverse ambitious fragmented goals but not to solve any single grand deep mystery. On the other hand, their advances enable solutions to the other deep mysteries, such as building cameras capable of seeing the furthest galaxies or the smallest cell scaffolding.

This fragmentation of the raison d'être of constructors is what makes them both more invisible but more pervasive than simplifiers. In physics we might take the paradigm of the hunt for the Higgs particle as archetypal simplifier science (figure 2.4). A similar scale of effort for constructors would be the vast group of people trying to understand the implications of Maxwell's electromagnetic equations for the last century. Recent fashions (known in science as *bandwagons*) have shown how little understanding we have for the inherent potentialities hidden in such equations of light waves, despite our knowing exactly how to write them. We simply cannot

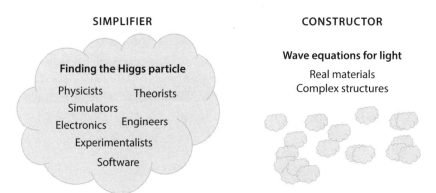

Figure 2.4: *Simplifiers* tend to congregate around grand challenges. *Constructors* explore research areas in many competing research teams, congregating fleetingly around hot topics.

predict what situations might be interesting to construct with intricate networks of submicroscopic metallic wires to make "metamaterials," illuminating what Abraham Flexner in 1939 referred to as "the usefulness of useless knowledge." Perhaps tens of thousands of scientists worldwide are chipping away at different parts of this challenge to understand the waves of light. But the scientific participants never think in terms of it being a universally motivated grand challenge, because they work in small groups competing with each other on tiny detailed parts of this whole. Intellectual fragmentation is the order of the day in constructor science.

Public perceptions of constructors have developed more recently and do not evoke the same distinct emotional response as simplifiers. Later we will discuss their relation to engineering and technology, but they often appear to be involved in an abstruse and complicated problem that is distanced from us by unfamiliarity. It is not always easy to identify a clear visionary goal for a constructor: just as buildings can always support a new extension and cities a new suburb, small features accrete. But, out of the whole process of constructor science emerge new ideas and technologies that pervade society. Their creative process emerges not just from one individual or a specific research group, but often from a large network of cooperating and competing research teams. Depicting such a story can be very complex and is rarely attempted in its breadth. We are mostly thankful for the technological advances as they come into our hands, but unclear and disconnected from what science made them and when. We also have an ambivalence about the uses to which this science is put, from weaponry to resource extraction, which is different from merely knowing the simplifying answer to a scientific question. When we give scientists the responsibility for their offspring, is it the initial simplifiers or the following constructors who we mean to implicate?

These roles of simplifier and constructor scientists will be useful throughout this book, such as in considering why so much media and political attention is focused on simplifiers. We will see however that many more researchers identify with being constructors. This disconnect can be understood better through the ecosystem of science. Before I discuss this though, we need some better idea of how many scientists in the world there really are.

HOW MANY SCIENTISTS ARE THERE?

WHAT COUNTS?

The throwaway answer to the question of how many scientists there are, is "over six billion," given the in-built propensity for *every* human, from their earliest childhood, to express curiosity about their surroundings through experiment and musings. Toddlers have been clearly shown to make hypotheses about their physical observations, test them, and come up with theories and beliefs. Despite the complaint that our school education systems seems to squeeze out this innate curiosity, I really don't believe it. Challenging my friends, young or old, with puzzles from my hoard of brainteasers such as three-dimensional interlocking "burrs" or undulating arrays of poised magnets, and they *all* tend to interact with them in ways that precisely mirror scientific research. It is just that only a fraction of our societies are formally employed to use these talents on science itself.

The global scale of professional scientific research can be recognized by estimating how many scientists there are, and how this number might be changing. Most scientists themselves, while they have a sense of the researchers in their own subfield and a strong sense that the number they have to deal with is surging, have only a hazy idea of the total number of scientists burgeoning around the globe. The scale of the community as a whole is not in their grasp.

The most recent estimates produced by UNESCO (from national governments' data) suggest that the number of professional scientists in the world is over eight million. This was up from five million in 2002, to the most recent data from 2013 giving nearly eight million. That is a lot of research projects, each of the eight million involved in threading a gap in the web of knowledge. On first encounter this number seems frighteningly large to me, especially since they are highly concentrated within small parts of the planet. Imagine a city on the scale of London, Cairo, Moscow, or Beijing, with each occupant a scientist and all vying for recognition, intensely involved in their profession. This is a very large number of people doing research, and also a staggering rate of increase. Already in 1961 Derek de Solla Price studied several measures of science activity and showed them all growing at an exponential rate. One of his conclusions was that more than 90 percent of all scientists who have ever lived are alive today, and this appears to have been true at every point for nearly three hundred years. But things are changing.

THE MULTIPLICATION OF LAB COATS

Clearly eight million scientists cannot be aware of each other's daily work, so science is necessarily fragmented. Where are they all? In most of the Western nations, the fraction of the population counted as being in research and development (R&D) consistently settles around 0.4 percent. We will discuss definitions later, but essentially all these people are involved in finding out new things. So far there are many fewer scientists than this in China (0.1 percent), India, and all Africa (0.01 percent). One implication is that as these nations industrialize and develop, they will focus similar fractions of their human resource on science, and the total number of scientists worldwide will exceed twenty-five million, a further tripling in the next fifty or more years. It is unclear whether they will enter existing subfields, or the branches of science will multiply. Perhaps the existing concentration of scientists will just spread out more evenly with decreasing numbers in Western countries. Perhaps the pace of science research will increase, and we can look forward to reading of more science breakthroughs each year, or maybe this is already at the limit. Will the quality of research they produce stay the same, or will soon each scientist be allowed to devote longer spans to getting it right? As we will see later on, it is the internal workings of the science ecosystem that give some indication of how these matters will be resolved.

Understanding the numbers is useful to grapple with such questions, and many more to be raised. On the other hand, I want to avoid burdening our discussions with roving packs of figures since this is unlikely to smooth your path to a holistic portrait of our science world. My strategy is going to be to give general messages in the text, but allow you to look at the raw data yourself as much as possible. The blog and website that go along with this book (thesciencemonster.com) have as many as possible of the links and raw data and some of the analyses. Because much of the data needed to get a fuller picture shuffles in the shadows of obscurity, a developing online community is needed to improve it. For now, inspecting the sweep of statistics will let us spot the diversity between countries and emphasize regional changes underway.

One aspect of interest is how fast the number of researchers on the planet is increasing. The average annual growth rate of scientists is now over 4 percent, in contrast with the annual world population increase of only 1.1 percent. This is a massive expansion of science that would lead to a doubling of researchers every sixteen years. Although the rate globally is 4 percent per year (figure 2.5), it is much lower in the Western developed nations of North America (1.6 percent), the EU (1.5 percent), and Japan (1.9 percent), all of which expanded their science base as early as the 1950s. More recently China (12 percent), India (6 percent), Brazil (12 percent), Korea (9 percent), and lately Russia have been drawn to imitate the Western research model based on university and government investment, with rapidly expanding numbers. These nations have been convinced that science research underpins their economic and manufacturing expansion in high-technology areas. Inverting this view of cause and effect would instead suggest that any nation with a massively expanding high-technology industry supports a new elite who believe in sustaining science research.

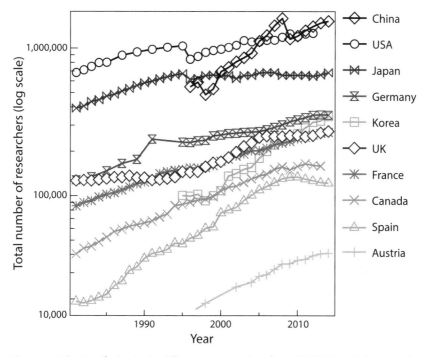

Figure 2.5: The rise of scientists in different countries. Data from UNESCO, includes researchers in universities, institutes, and industry.

WHO IS A SCIENTIST?

In considering these trends, we should be cautious of simply accepting statistics from UNESCO without understanding the difficulties in such comparisons. Only recently have agreed measures of who should be included as a researcher been adopted by different countries. These "Frascati" definitions dating from 2002 and used by UNESCO, define R&D as "creative work undertaken on a systematic basis to increase the stock of knowledge, and its use to devise new applications." It is distinguished by the presence of an "appreciable element of novelty," so it excludes routine testing, market research, patent applications, or trial production runs but includes science development. This seems to me to correctly capture the wider community of scientists and the shared way that they work, combining researchers from industry, higher education, and government institutes (normally only a small fraction).

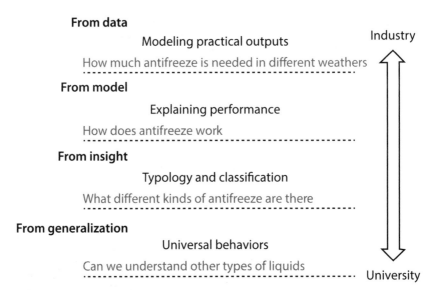

Figure 2.6: How different focus on the same problem can support technology and basic curiosity.

To see the richness of approaches to science more clearly, I will imagine a situation that uses real data to identify optimal performance properties, for instance the formulation of antifreeze to add to a car engine coolant system in different seasons (figure 2.6). One type of approach to "exploring" this might be to find the exact mixtures that seem to be useful (modeling practical outputs). A slightly more general approach instead is to find out why it works this way, producing a model dependent on the particular antifreeze properties (explaining performance). Another type of approach asks how different behaviors might arise and under what conditions, predicting performance of different types of cooling fluids (typology and classification). A further level might be to develop the thermodynamics of liquids when heated and cooled in pressurized flows, encompassing many types of liquid effects, and considering whether these can be seen in nature, or can be created in a lab (universal behaviors). All these studies have both clear questions and answers and divide between pure science, applied science, engineering, or technology. They also span theory and experiment. Some answers have only a limited region of applicability, perhaps of interest for a particular company honing its production parameters. Other answers might provide an entirely new way for a scientific field to organize its observations. And these different types of science emerge from the same problem, so that a researcher might pass rapidly through all these disciplines by slightly changing the details of their inquiry. As long as new knowledge is built by creative systematic novel inquiry, it seems right to me to call all of it science.

The boundary between engineering science done in a university environment and that done in a company is the degree of particularity—how focused the scientist is on a specific aspect. Most researchers I have known in companies bemoan the fact that they never are allowed the time to understand things properly—they are just required to fix a problem as quickly as possible, provide the pertinent data, and move on to the next issue. In some cases, they aren't creating new knowledge so I wouldn't consider what they are doing to be science. University researchers by contrast are asked to understand a problem more abstractly, providing knowledge that is reusable in other contexts. Flows of money back and forth between the different funders mix up these compartments. Since technology is able to head onward even when being based only on trial-and-error advances, its underpinning science understanding can often be far behind, and this catch-up has been one of the drivers of science for over five hundred years.

The Frascati definition also insists that postgraduate students learning to do research (the apprentices in the guilds of science) should be considered to be researchers, expanding the higher education totals far more than the number of permanent academic scientists. In the UK for instance, there are only thirty-one thousand or 0.05 percent of the UK population who have *permanent* university jobs in the sciences, medicine, or engineering, which is 20 percent of the total researchers in higher education, and 12 percent of the total number of UK researchers reported to UNESCO. For comparison in the United States in 2010 there were 270,000 academics or 0.09 percent of the population with science and engineering faculty jobs, about 23 percent of the total number of researchers. Which types of researchers are growing the most varies between states: permanent academics (China, India), researchers in industry (Korea, Spain), or more short-term contract researchers (UK, the United States). The fraction of scientists who are in universities or institutes in each country (as we will see in chapter 3) varies from over 60 percent in Spain or the UK, to below 25 percent in the USA and Korea, reflecting how much businesses are able or willing to invest in research compared to their governments.

What controls the number of scientists supported in each society? Given that scientists encourage the interests, training, and funding of ever multiplying generations of younger followers, then fecundity is built into the science ecosystem just as in its biological progenitor. The constraints as in the natural world are the resources that underpin this growth: its funding and the size of its habitat. Health care in each country is funded about ten times more than all publicly funded science research, and the inexorable rise in this cost of preserving health is causing worries everywhere. By comparison science research is a small but noticeable fraction of a society's budget, stabilized at the level about which taxpayers and civil servants are probably comfortable. Science spending seems to rise until it becomes noticed and then stops. But there is nothing well defined about this level, no formula that says funding should not be halved or doubled. How much science is the right amount of science?

Industries employ various numbers of (mostly constructor) scientists to help innovate their products or services. In this constant struggle to generate continued and growing revenues to stay alive, scientific research is one of their drivers. But too much research sucks resources out of other parts of the business and might not pay off, so the balance is delicate. The number of scientists supported by business has been increasing in the last decades. In most cases this rate matches the increase in government-supported researchers (excepting only the UK, where their number has not increased for twenty years). So it seems the increasing wealth of societies becomes mirrored in support both for publically funded science, and for the opportunities companies have for investing to drive forward their technology base.

The number of scientists cheerfully supported by a society in its institutes and universities depends on another aspect of their jobs as well, the training of young undergraduate students. Universities are historically organized into disciple-"owned" departments to do this teaching. The expansion in the number of young people going on to higher education in many countries has justified the rise in the number of their professors. But when science is perceived as a difficult subject to excel at, and there is high promise of well-paid jobs in other sectors requiring numeracy (such as finance), the numbers of students studying science drops. If these perceptions reverse, science student numbers rise. When many fewer study within a department the financial pressures build up on administrators, passed on from government departments. Several times in recent decades there has been concern with students in a specific country fleeing science, and sometimes academics have lost their jobs. A slow rebalancing then comes through the delayed connection to the economic situation of a country, which eventually influences the demand for the number of science educators. These dawdling tugs from the economy also mean that the current status quo rarely satisfies anybody—there are always perceived pressures to change. But the inexorable increase in science students worldwide supports the increase in number of scientists.

Where do new academics go? What fields of science do these permanently employed researchers ally themselves with? Because fields of science are fractured and split across different departments in different countries, such comparisons are incomplete. In the UK a series of government-mandated research assessments provide some useful data, splitting the thirty-one thousand academics into 25 percent physical scientists, 16 percent biological scientists, 34 percent medical scientists, and 25 percent engineers. The United States has a similar balance among tenured academics with 30 percent physical scientists, 52 percent biological or medical, and 16 percent in engineering. In the last decade, the dominant increase has been in the number of bioscientists, so they now outnumber their physical colleagues nearly twofold. While this balance is different in different countries that prioritize different fields, there is no agreement, or even discussion, about what would be the right balance. The juggernaut of opportunity rolls on, and different disciplinary groups manage to fleetingly sway the ear of politicians.

The world growth rate in researchers of 4.4 percent corresponds to over three hundred thousand new researchers each year. In the

United States there are nearly twenty thousand extra new researchers in industry and five thousand extra university researchers each
year. In China 190,000 new researchers are employed each year, over
150,000 in industry, and more than 20,000 in academia or institutes.
Worldwide, more than ninety thousand university or government
researchers are added each year, the vast majority in science, health
and technology. That is on top of two million such researchers already and doesn't count all the technicians and support employees
that make their research possible. What do all these researchers do?
Does the world need this many new scientists? How do they find
new research fields? As we survey the science ecosystem, we will later
see that instead of sparking entirely new fields, most of these people
follow existing paths because these give them greater credit to benefit their careers.

HOW SCIENTISTS SITUATE THEMSELVES

The link between asking questions of the natural world, and un-
earthing answers that allow technologies to be built, has produced
a twin cycle of evolving science and evolving technology. By now
these are so entangled that every scientist has a stake in both posing
new questions, and working forward the implications from previous
answers. However, science is so large that there is no possibility of
remaining equally involved in everything. All human society abhors
homogeneity. The emerging segregation of the sciences has given us
divisions into disciplines, domains, and paymasters, as well as between countries.

Disciplinary boundaries have become immovable. Since natural
science expanded in the Renaissance and then exploded with industrial Victorian society, it has been fragmented into disciplines in which
different natural laws take center stage. Despite their underlying connectivity, the disciplinary compartments divide science into defined
perspectives and levels of scale. Chemists use restricted packages of
the truths abstracted from quantum physics to explain the bonding
of different atoms into molecules, and their subsequent reactivities.
Geologists encapsulate the chemistry of mineral and biological cycles
to help them understand the persistence of deep slow time on shap-
ing our bedrock and boundaries. Discipline barriers partly help in

training scientists, who can concentrate on a well-defined part of the whole massive structure and learning its foundations, allowing themselves to be drawn to areas that intrigue them most. Despite many discussions about whether this is the best way to train anyone, it has become a firmly established component in the science ecosystem. In terms once described to me when working at IBM, you can order up "one chemist" and know the sort of scientist who is going to turn up. Beyond just defining a skill set, discipline training also defines perspectives across the web of science knowledge.

The simplifier/constructor domains knit together disciplines that have common styles and foci. The natural sciences typically support the more simplifier sciences: mathematics, physics, chemistry, biochemistry, earth sciences—disciplines that provide many of the underlying principles for other domains. Engineering uses these for constructor science but tries to abstract the principles for best outcomes. Medical sciences and health care combine both simplifier and constructor science, focusing on questions and answers of the living world from molecules to mass inoculation. We should also note that included in UNESCO totals are other much smaller categories: social scientists who study systemic properties of economies, of people, or of networks; agricultural scientists who were already students of the natural world ten thousand years ago; and (comparatively far fewer) researchers in the humanities.

To get a historical view of the simplifier/constructor balance is not easy. Their relative populations seem not to change dramatically, but this is hard to track. One aspect we can follow is the evolving split between academic and industrial researchers. Data from UNESCO shows that the fraction of researchers working inside businesses *within* each country has changed by less than 20 percent over the past thirty years, which is far less than the 60 percent difference *between* countries in their fraction of industrial scientists. Different societies seem to support a different mix of science between public and private spheres. But this stability clearly shows it is not that engineering has taken over with the rise of science-based industries—academic research has kept pace with the expansion of technology. Both simplifiers and constructors are expanding, in both academic and commercial domains.

One final consideration dividing hierarchies of scientists is who pays them. More than 60 percent of the eight million R&D scientists are in industry, where by and large they have to concentrate on science that advances technologies that we consumers will buy. While we depict university scientists spending their time trying to find and understand the building blocks making up the edifice of science, checking which components are weak or flawed and might fall down, instead industrial scientists spend their time honing, polishing, and gluing the blocks that must not be allowed to fall apart.

The discussion of who contributes what to the different components of this science edifice is rooted in human tribal societies. Despite the guise of rational enthusiasms among scientists, there is a great deal of self-reinforcement and group tribalism about who is more important, and many attempts to impose a pecking order. Since choosing any one definition of science sets who is top dog, this is far more than a play on words, but is crucial for who can amass resources to generate more impact, in a self-reinforcing circle fought in the arena of the science ecosystem.

As I described above, the simplifier/constructor split is not the same dichotomy as that of science/engineering, pure/applied, or theory/experiment. Constructor science can often be found in engineering departments in the United States or Japan, while in the UK device innovation is often pushed within physics or chemistry departments. People call themselves scientists or engineers depending on perceived status within the cultures in which they were educated, or the culture in which they now work. As with all stereotyping, the boundaries become more blurred the closer you focus on them. An engineering academic might be asking fundamental questions about limits to performance, a mathematician might be devising practical cryptographic methods for the Internet. While engineers are often constructors, many more constructors would avoid calling themselves engineers. Constructors are scattered far and wide across science, wherever answers to questions suggest creativity that can be wrought by humans.

THE VIEW FROM OUTSIDE

In some academic settings the purer the scientist the higher the esteem, while in companies the reverse can be true. But what do the public feel makes up the pecking order of science? The internal conceits of the science community seem to have leaked into a wider framing, so that the postures above may sound already familiar. A general perception of science is that it is more to do with the fundamental explorations by the simplifiers. Somehow what emerges out of this are technologies that we value and have even come to expect. How the two are connected is rather hazy in the public mind. Constructors are lost to view. Consider the reputations of Isaac Newton (simplifier) with his contemporary Abraham Darby (constructor). In modern times we venerate technology visionaries and integrators (Steve Jobs, Bill Gates) rather than the constructor scientists who enable their technologies. Even students coming into the academic disciplines have these attitudes, attracted as they are by "explore the world" refrains, promotional slogans competing for their hearts and minds. Only after a few years of study, in which they see the romantic carapace exchanged for a messy interior, does the attraction of science's utility start to gleam for this younger generation (perhaps equally misleadingly). The middle ground is strangely absent from all these perspectives, and where characters appear (for instance Nikola Tesla or Alan Turing) there is a diffuse haziness about their real contributions.

The public sees science as what simplifiers do, and technology as what engineers do, but has little idea of what constructor science might be. Constructors however see themselves clearly as scientists, motivated both by the urge to build new things to understand, as well as by the beauty of their creations. There are at least as many of them as simplifiers, but they are less organized around grand challenges, and so flow like grains of sand through the fingers of society rather than weigh in its hand as a solid rock. Simplifiers like to feel they are the "real" scientists, but in explaining the need for society to invest large sums in them, they have learned to point to the potential spin-offs that come from their work. These spin-offs in fact mostly arise from intermediary constructor science. Such tensions between simplifiers and constructors affect how vast sums are spent. Understanding this tangle allows us to better evaluate the demands from these different communities on our own society's resources.

Another way to approach the question explored in this chapter, "What is science?," is to see it holistically as an ecosystem. I use this metaphor throughout the book as a vehicle for surveying and understanding the science enterprise. So now I will introduce some of the ideas behind ecosystems, and make tentative connections to the organization of science. In later chapters, after examining the realities of science, we will return again to look back at this holistic ecosystem view.

WHAT IS AN ECOSYSTEM?

Ecologists study the arrangements and abundances of living things and how these emerge and evolve, through their relations and exchanges. Drawing the dividing line between what is inside or outside any ecosystem is a little arbitrary, but the easier it is, the better defined an ecosystem is resolved and the more sensible its study becomes. In bio-ecosystems, the way that both energy and nutrients cycle through the myriad parts is crucial. Almost all life is supported by the rays of sunlight sweeping the landscape, absorbed in the magical green molecules of algae and plants, and passed step by step across the entire spectrum of life.

To put together the equivalents in our science ecosystem, we have to amass all the actors involved. Beyond the massed tribes of scientists, we should include the funding councils, science journals, media for science such as magazines or blogs, and the learned scientific societies, as well as the institutions where science is done, and embracing also within this tribe many research-rich companies (figure 2.7). Scientists themselves are both inside and outside this ecosystem as they pay taxes, support charities, and buy products that support science. The professional part of their lives though is firmly planted within the ecosystem, separated from the wider public. Our actors, whatever their size, weight, and constitution, all perform as if they were living organisms and make choices depending on the influences and interactions within their local neighborhood. Some influences are close by in the virtual landscape of science, and some influences are very indirect, passing through clouds of intermediaries. Drawing the envelope around the entire science ecosystem in this way produces a boundary that is not too porous, with flows inward and outward that make intuitive sense, and possessing internal dynamics that can be scanned.

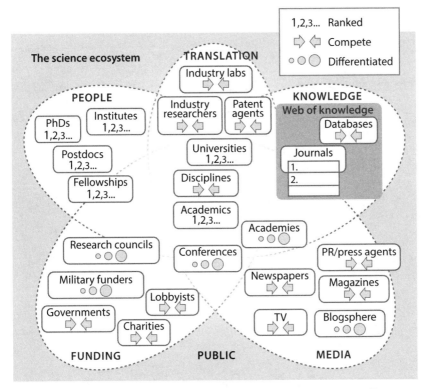

Figure 2.7: The science ecosystem actors. Each species component competes internally. In some cases a *ranking* within the actors of a species is very evident ("1,2,3 . . .") through metrics, such as for individual universities. Other species interact in strong *competition* (arrows), but it is not clear who is on top, such as lobbyists. A further set of species are more *differentiated* so compete less directly (circles), such as research funders in different areas of engineering or medicine.

THE SUNLIGHT OF SCIENCE

If solar energy is at the root of biology, is money at the root of science? Certainly it is the lowest common denominator for anything to happen, from the roots supporting the science research itself, to building the infrastructure, feeding the foliage of journals and media, and training budding scientists. As a basic input then we might use "funds" as the equivalent of "sunlight," an essential power source but not the only input (figure 2.8).

Harvesting sunlight at the bottom of the food chain filters a free resource bestowed by the munificent star that we orbit. However our plants and algae compete intensely to harvest the greatest solar

THE NATURAL WORLD

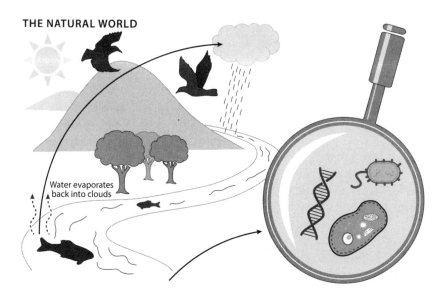

Water evaporates
back into clouds

THE WORLD OF SCIENCE

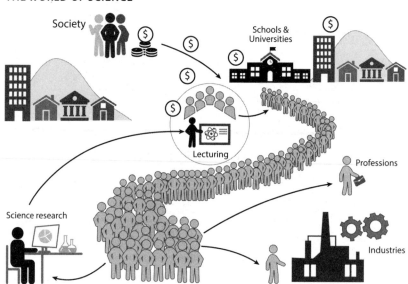

Society

Schools &
Universities

Lecturing

Professions

Science research

Industries

Figure 2.8: Ecosystems of nature and science. Cycles of water, and of people.

bounty, stealing light from each other and in the process creating forest canopies or vast ocean blooms that arise from and then influence
evolution. While money is not showered over the planetary surface,
in our large societies it is widely available for science from taxes,
profits, investments, and donations. These crucial inputs to the sci-
ecosystem fluctuate more than the few percent oscillations caused
by our solar orbit. Globally, the raw input of money to science has
continually increased over the last few centuries.

RESOURCES OF THE ECOSYSTEM

In bio-ecosystems cycling of nutrients is profoundly important, lim-
iting and sculpting the flora and fauna of each region. The crucial
bio-geo-chemical cycles of water, carbon, oxygen, nitrogen, and phos-
phorous depend on geology and biochemistry at molecular and
mountainous scales. Scarcity of any of these components greatly
hampers the lushness and range of life, abandoning it to niche species that have specialist adaptions to harsh conditions. Organisms
that accumulate the most of these resources will thrive best, leading
to intense competition for nutrients.

Mimicking nutrients in the sci-ecosystem are different resources:
people, infrastructures, and formal structures (such as well-defined
career paths, or regular conferences), which I consider in turn.

People are the lifeblood of science (figure 2.8), emerging from
the avid interests of youngsters, through the training of graduates in
science subjects, to researchers, presenters, writers, filmmakers, editors, referees, conference organizers, committee members, and ever
on. This raw human material goes on to directly fashion specific actors, as well as to populate the institutes and organizations that form
other actors.

Infrastructure is bequeathed to us from the past, forming a landscape of universities, synchrotrons, telescopes, microscopes, and databases that allows science to build on its successes. The material side
of this is physically fixed, sprinkled unevenly over the surface of the
planet, and resistant to rapid change. The electronic infrastructure
may be globally accessible but it is locally embedded inside institutions and institutes. People can move around but are rationed by
their costs and by their skills.

Formal structures are also bequeathed to us from the past, as the accepted know-how for making systems work. For instance, to become a scientist there is a well-defined route to follow, which has settled into a globally accepted norm—the criteria for the PhD apprenticeship was set by guilds from the Middle Ages. The system of publishing science follows a peer-review system that was similarly set by natural philosophers after the emergence of formalized curiosity in the Enlightenment, as we will see in chapter 4. Such mechanisms, unlike nutrients, are not limited resources but can be freely cloned. Like genes that jump species and spread useful adaptions more widely across organisms, these parcels of helpful tools tend to evolve in their own right as packaged ideas or memory-based concepts often called "memes." Each unit in the science ecosystem is buttressed by a collection of memes (its "memome") that have settled to cohabit comfortably, ejecting ideas that cannot fit in easily.

On the other hand some things cannot simply be cloned. The number of science degrees available to take in each country is set by how much each society devotes to educating and training in science. Limited by money, it becomes a bottleneck of the science ecosystem, constricting the pool of potential scientists to be taken up. Such flows of nutrients are critical to an ecosystem, creating and cycling resources.

THE HABITAT OF SCIENCE

The physical conditions that surround a community, including the land, air, and water as well as external influences such as climate and topography, form its habitat. Institutions are a strong feature of the science habitat, with many scientists embedded in universities or vast companies. Just as the characteristics of rock set the plant species that can thrive on them, scientists find the nature of some universities more conducive to productive research than others.

Increasing global competition for identity has made universities much more aware and responsive to each other. Previously, diverse institutional microhabitats supported a profusion of different attitudes and spanned many ways to support research. Larger-scale habitats can form in different countries, encompassing different styles of university, academic career, funding mechanisms, or public esteem and visibility. However in recent decades it matters less *where* across the world that a scientist's university is located. In the same globalization spirit, companies have become much more similar in the attitudes they hold toward their research labs, erasing distinct cultural discontinuities for instance between Japanese and American industrial innovation. Habitat diversity has reduced, as habitat coalescence has sped up. The way that science is done in different regions is far closer than in previous epochs.

The political climate dramatically alters freedoms and funding, determining when, where, and how much is driven into different parts of science. Conversely on the rapid timescales of weather, all the metaphors we are accustomed to, spanning droughts, storms, tornadoes, and irrigation, have become appropriate epithets within the science policy climate. Political ideas are only slowly changed by the progression of science itself. Climate change is a topical example for the science ecosystem, with long-term predictions only gradually changing the terms of political debate and action following far behind. Mostly then, "political climates" are *external factors* to doing science, and we do not have to include mainstream politics inside our science ecosystem to encapsulate most interactions. The evolution of rationales for investment in science frames the discussions about what science to do now, and also what science to highlight that has been done.

SCIENCE ECOSYSTEM SERVICES

We depend on both the goods and services of ecosystems. *Ecosystem goods* are their tangible outputs, such as practical materials like food, wood, or oil. *Ecosystem services* are less concrete improvements, which include maintaining oxygen in the atmosphere, cleaning air and water, or crop pollination. They also include even less tangible aspects such as the beauty and inspiration of wild moorland, canopied forests, and high deserts. One of the major problems brought about by capitalism has been the consistent undervaluing of ecosystem services, which are taken for granted until eroded into collapse.

Applying these terms to the science ecosystem separates tangible outputs, such as new technologies, improved medicine, and the training of skilled and numerate people, from the less tangible outputs. Services from the science ecosystem include improved understanding

NATURAL | SCIENCE
ECOSYSTEM

Sun	£, $, €	Climate	Politics
Rain	People	Weather	Science trends or bandwagons
Soil	Freedom	Habitats	Institutions
Nutrients	Infrastructure	Biodiversity	Diversity

Carbon fixing	Knowledge fixing	Ecosystem goods	Technology, science, training
Decay recycles	Change recycles	Ecosystem services	Beauty, understanding, inspiration
Genes	Memes	Resilience	Inertia, peer review
Adaptations	Brands	Resistance	Institutionalized

Figure 2.9: Ecosystem translations from natural world (*left columns*) to science world (*right*).

of our world, our planet, and the frameworks of the science, leading toward rational approaches to societal and political problems, as well as opportunities to level inequality between different countries. But it also includes many of the things we value about science but have found hard to articulate: the beauty of mathematics, a deep motivation for the inherent values of education, the intrinsic worth of transnational scientific communities, and the value of scholarly discussion within academic collectives. Such *science ecosystem services*, which we can term *SciES* ("sighs"), have been undervalued and often taken for granted by funders and governments (figure 2.9).

STABILITY AND CHANGE

One reason for mankind's increasing focus on ecology is the visible changes wrought by ourselves on the biosphere of our planet. Researchers have highlighted ecosystems that are vulnerable and tried to understand why this is. A stable ecosystem is *resistant* when perturbations do not shift its balance much, and *resilient* when it returns speedily to the status quo. On these technical definitions, science ecosystems are generally very resistant to change, but their resilience varies widely across disciplines. Physicists move subfields quickly when pushed or pulled: they have constantly been tempted by new funding and intellectual intrigue into areas such as engineering, chemistry, ecology, biology, and neuroscience. Material scientists in different countries have responded to the same temptations at very different rates, expanding their interests more rapidly in the United

States than in many cultures. By contrast, the distinctive culture of chemistry has made it rather slower to evolve outside the divisions made over a century ago into formalized branches of inorganic, organic, and physical chemists, which still more rarely collide.

This resistance and resilience to change sets much of the character of science. Despite scientists' avowed interest in their own internally driven curiosities, they respond quite rapidly to sticks and carrots, mostly because they make crucial decisions on their own, rather than collectively. On the other hand, the funders, publishers, learned societies, conference committees, and other organizing actors respond very slowly because they are collectives and acquire their resources in much less diverse or explicitly competitive ways. The science ecosystem is thus built from components with very different timescales. To individual scientists, their elite institutions appear to move glacially slowly, leading to frustration or contempt. It can be similar within large companies. This is one origin of the common feeling of powerlessness for scientists and their compatriots, of being trapped in the system.

DIVERSITY

Biodiversity changes constantly. It decreases when an invading species without natural predators outcompetes everything and takes over. Disease can also devastate biodiversity when interbreeding has reduced the gene pool so that entire populations are susceptible to catastrophic infection. Paradoxically, high rates of evolution can also reduce biodiversity because closely akin species do not develop radical new modes of life but simply compete with each other on the same terms. Diversity is crucial for resistance and resilience.

DARWINIAN COMPETITIONS

My motivation in this book is to describe the human-constructed world of scientific research in terms of a restlessly evolving ecological system, but we should question where this analogy might mislead. Just as in nature, competition is at the root of many features of this system of science, in the sense that only the most successful parts survive and expand. The evolution of life is based on two key ideas working together: "survival of the fittest" and "descent with modification."

The first competitive aspect is provided by any collection of autonomous agents with different strengths and weaknesses, interacting within a landscape of rationed nourishment. The second plank rests on internal blueprints that explicitly create these agents. Offspring will be almost the same, but mutations and sex create modifications, which accumulate over generations whenever they help survival.

We don't have blueprints for scientists. Their children don't research their same field (though their progeny are more likely to become scientists). Yet there are analogies with genetic inheritance. Scientists do teach young researchers in their labs, who retain their style, manner, and research interests but subtly reconfigured. Journals produce offspring in their likeness. Funding bodies get wound up, split, or reconfigured, often with many of the same staff who convey a culture with them. Even without genes for institutions that would guide how these should be built, their descendants are formed almost entirely in the same institutional mould. At the heart of this mimicry of phenotype is the conservative nature of human organization. Despite the lack of formal blueprints, we carry with us ideas of how something should be done. We find it hard to break from our experiences. This makes it doubly hard for scientists whose main challenge is to develop new ideas.

Reprising then what science is, the Frascati definition I pointed to nicely emphasizes novel, systematic, creative work increasing scientific knowledge. The total number of scientists under this definition has been continually increasing and looks set to further triple before the next century if the world continues to emphasize the education of its youth. I showed that science is done both by simplifiers and constructors (but more of the latter), who have different focus and aims. I also introduced the larger array of people, institutions, and actors who are make up the interacting world of science. I presented a useful metaphor to view its workings as an ecosystem with many competing components, which gives rise to concepts that will be useful later. I now start our exploration of the ecosystem by inquiring first what motivates scientists and why they might be beneficial for our societies.

MOTIVATING SCIENCE

What is the point of science?

The motivations for doing science vary for each scientist, depending on their situation within the ecosystem. What will become evident in this chapter is how much science is initially pursued without a societal end purpose in sight. Curiosity is most often the driver, and yet from this emerges a utility that yields new technologies that allow yet more science to be done, in a virtuous circle bridging the full science spectrum.

In this chapter I start by looking at how the different roles of scientists as simplifiers or constructors vary between countries. Then, to ground a discussion about how curiosity drives research, I analyze one of the major accolades of science, the Nobel Prizes, for insights about motivations and utility.

ARE THERE MORE SIMPLIFIERS OR CONSTRUCTORS

Simplifiers want to understand the world's natural scientific system. Constructors want to use the insights to synthesize new scientific domains. To gain a perspective on the science ecosystem, it is helpful to see which are supported. While the world outside science assumes scientists mostly fit the stereotype of a simplifier, are there actually more constructors? Counting precisely is simply unachievable, in part because each individual scientist may at different times in their life be motivated by different questions, taking on these different roles at different stages. Even within the same piece of research, understanding can rapidly give way to its exploitation. The reverse journey is also prevalent, in which a constructed science field with rich and emergent properties becomes the major focus of simplifiers who try to understand it more deeply.

Sometimes scientists can look at the summary of a piece of research and rapidly decide which role it fits into, but this is a very hard task to automate. As the most crude gauge we can assume that scientists in industry are mostly constructors, trying to create something new. The fraction of industrial-based scientists varies from 80 percent in the United States, to 35 percent in Spain or the UK (figure 3.1) and is about 61 percent averaged across the world. It seems then that over half of all scientists are constructors.

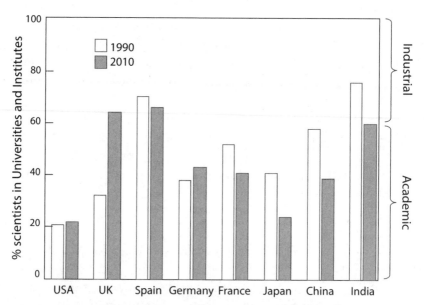

Figure 3.1: Fraction of scientists who are academics (bars, *lower*) or industrialists (remainder, *upper*) in several countries in 1990 and 2010.

To understand how much more than half, we need to understand the split between simplifiers and constructors in universities, but there is no easy trick to estimate this. We might look at the numbers of academics in each type of department, and try to assign the disciplinary departments wholesale to one or other focus. Or we might take each scientist and use the research journals that they publish in to judge which type of science that they do (or even what fraction of their work is aligned along simplifier principles). However as we shall see, many journals publish both types of science alongside each other within the short reports called "papers," so we would have to find a way to look at each and evaluate what question motivates the work: unpicking which is an exploration of nature, and which instead aims to devise new creations.

THE NOBEL GAUGE

One manageable approach is to sample a subset across much of science that is small enough to be examined in detail. I have chosen here to see if an analysis of the Nobel Prizes in Physics, Chemistry, and Medicine and Physiology over the last sixty years can tell us much about the changing face of science. As with the rest of the analyses in this book, the data is online at thesciencemonster.com for you to take your own look. Despite the snag that they identify only a few individuals in a collective enterprise, Nobels highlight moments in the continuous narrative and identify rewards from steadfast investment (figure 3.2).

I will first say straight off that I am not a fan of the Nobel Prizes as a motivating influence for scientists, because this accolade depends on recognizing *the* scientist who has "conferred the greatest benefit on mankind" in a discipline. It frequently requires sharp elbows to draw clear water between an aspiring sole discoverer and the entire remainder of scientists in the same subfield. More problematic to my view is that the science enterprise is collective and connected, and even the most brilliant leader relies on the many prior advances or struggles of others. Some active fronts of science do not have three clearly isolatable figures who are seen to have initiated it and can carry the Nobel accolade for their subfield (at most three people

Figure 3.2: Nobel Prize ceremony picks out each year a maximum of three individuals who were identified with a key advance in medicine/physiology (*top right*), physics, or chemistry (*bottom right*). Image credit: Nobel Media AB 2016/Alexander Mahmoud.

can share each science prize). So within science I see the Nobels as highlighting individuals at the expense of the whole, highlighting moments out of a continuous story, as well as highlighting easy narratives at the expense of complicated reality.

On the other hand, one of the great benefits of Nobels is that they highlight to everyone some really interesting sites of active science, and show a reward for steadfast investment made many years ago—on average Nobels come twenty years after a breakthrough, by which time a community has gotten over the shock and really worked out if it is useful and taken it up (or not). The fascination with individuals and narratives that is felt vital to dramatize every science story is heightened by the cult of top dog, the successful winner who has beaten all comers. Such exposure can enhance a thoughtful connection between previous societal support for research and what emerges in the longer term. So Nobel prizes are an ecosystem service that we would be poorer without.

The benefit of analyzing the entire cast of Nobel Prizes is that it assembles a parade of important research findings in diverse fields over a consistent length of time. So I will mine the last sixty years of the Physics, Chemistry and Physiology/Medicine Nobel Prizes to see what they say about the motivations of scientists. You might sympathize that it is not an easy challenge to read hundreds of summaries of breakthrough science in areas quite outside my own specialty so you should assume that some mandatory superficiality was needed to digest the results. However, it has been surprisingly possible to assign motivations for these original research projects, whether aimed at uncovering the scientific truth about something from the natural world, or creating something never found before in the universe. Only a few projects have elements of both (or have dubious assignation) so I regard these statistics as quite robust. Where prizes in a year were given for several different areas I treat them separately (hence I count more than sixty prizes in each field in these statistics). I decided to take prizes only after 1952, around the leap-off point for the modern era of science, and to compare the first thirty years after that with the most recent thirty years (which confers statistical errors of around 3 percent).

Within the Physics label (which actually includes many parts of engineering and material science), we find that simplifiers dominated three quarters of the Nobel Prizes from 1952 to 1981, but more recently constructors have edged the balance with more than half of those from 1982 to 2011 (figure 3.3). As the basic questions of how the world works are answered, physics has not stopped. Instead fertile realms have been opened up through creation of amazing new arenas with completely different properties, that are clearly human creations. The technology for enabling this has taken time to develop, growing through the second half of the twentieth century. It is likely that this trend will continue, with no cease in the richness of new domains in sight, but increasingly fewer being simplifiers (although still more prominently noticed).

In Chemistry (which spans widely from material science to biochemistry), a similar balance is seen in the early Nobels. Construction of new realms in chemistry, such as the long polymer molecules that make up rubbers and plastics, was established back in the nineteenth century, so simplifiers had their day earlier on. However the shift to constructor science is less abrupt, rising to only just over a third in the last thirty years. Looking at why this is, we see the recent rise of biochemistry that is still trying to understand how the microscopic

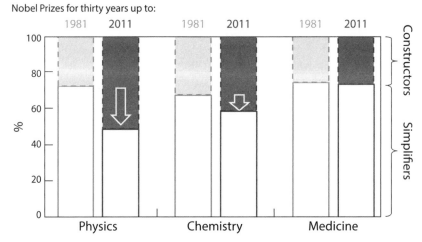

Figure 3.3: Fraction of Nobel Prizes that are simplifier or constructor, 1952–81, and 1982–2011.

machinery of life works at the most fundamental level. So a second wave of fundamental questions have been opened up, and simplifiers still have a long way to go.

In the domain of Physiology and Medicine, simplifiers are even more dominant. Almost nothing has changed in the last sixty years, with three quarters of prizes for simplifiers. There are as many puzzles about life from the cell to human disease as there ever were, and the opening up of genomics and proteomics has merely launched further questions.

Recognition takes time. The average wait from getting key research breakthroughs published to being wreathed in a Nobel crown was around fifteen years in all these disciplines but has been consistently getting longer, averaging now twenty-five years in the physical sciences, and twenty years in chemistry or medicine. The queue of worthy scientists has been increasing, but it also takes longer to understand what is a major breakthrough when the breadth of fields expands and it takes longer for discoveries to percolate across science. Overall in the 1990s, there were plenty of fundamental questions around in all science, which were still tractable enough to see major breakthroughs. The trend to note is that over time ever increasingly more constructor science commands the attention and respect of whole fields, as the ability of scientists to exert control at different size scales and of complex systems improves.

The rise of constructor science, and thus its increasing share of all that scientists do in the name of science, should I believe be more widely recognized. Even in the biological sciences, efforts to control, build, and create new fusions based on discoveries in the natural world have been increasing, and this will only intensify. To return to my question about balance, across all sciences one can use the current split between prizewinning permanent academics in different disciplines to estimate a 40:60 percent split between constructors and simplifiers. But this is even before factors that skew the recognition involved in major prizes are taken into account (with their greater emphasis toward fundamentals), and the time lag before any formal recognition emerges. The picture we should have in our minds for current scientists is the *creative constructive virtuoso* building something from nothing we have seen before. "More" can be radically different.

MAPPING THE STANCE

Early on in any field, when questions in the natural world are still fresh, most of the focus is on simplifier science. As time goes on, the fundamentals provided by nature become better understood, and if the knowledge can be taken onward, the trajectory of a field swings over toward constructors. Early simplifier science might enable huge advances in making utterly new and surprising systems, creating significant impact. Or simplifier science might not open up anything more than additional simplifier science, remaining limited to a niche area.

What might the map of such trajectories look like for different subjects? Most research affects the larger world of science in no major way. Some projects lead into directions that are hard to build on directly, as for recent string theories of particle physics. Some projects add a tiny bit of impact, improving our ability and understanding to build new constructs or techniques, nudging slightly constructorward. The very rare Nobel Prize type science can be situated anywhere but has high impact.

Strong science fields often move over time toward constructor science. But in turn this often delivers new technologies that allow the next steps in simplifier science to be taken. Work in the middle of the last century on superconductors (frigidly cooled metals that transport electrical current without loss), puzzling how these lossless states can arise, led to complex materials based on alloys and containing crucial levels of defects that supported extremely high magnetic fields. Making coils of these materials then became the basis of bending electromagnets that allow the high-energy particles at CERN in Geneva to be swung around, accelerated, and smashed together to reveal glints in the showers of new fundamentals. These electromagnets also drive the MRI scanners we routinely use to image deep inside our bodies and to map our minds. This is a typical cycle of high-impact science, spawning wide-ranging directions.

Similar stories have emerged from the molecules of life, with work on the unusual chemistry that holds them together, how they encode information, and how the chains are turned into useful proteins, all leading to the development of faster ways of decoding these sequences of information, now a major technology. This can then be widely used to understand simplifier science, such as how a single-cell egg develops into a trillion-cell mammal, or what the immune system is doing. The impact either way is clearly recognized by the scientists themselves, a selection of whom are asked to nominate and select the Nobel Prize winners. Scientists can be clearly appreciative of both simplifier and constructor science.

THE RISE AND FALL OF FIELDS

My scan through the Nobels allows us to capture the sway between different areas within each science prize. In Physics (and engineering), the first thirty years after 1950 were dominated by many discoveries of particle and atomic physics, which together took more than two-thirds of the awards. By the twenty-first century, all dropped significantly. At the same time astronomy has doubled its share, solid state (materials-based) prizes went up by more than half, and prizes in imaging such as scanning-tip or electron microscopies emerged.

Over the whole period the simplifiers sat predominantly in particle and astronomy fields. It is in atomic, optical, and solid state science that constructor physics has grown—fields that are termed "lab-based" science (composed of smaller teams working in many lounge-sized experimental rooms).

In the 1950s, almost half of the prizes were for theoretical advances, but more recently this has dropped below a quarter as startling results from experiments have had more impact. The evolving balance of academics in physics departments reflects this change with only a third being theorists now, at least in the United States. This rise of experimental physical science emerges from science fashions or "bandwagons," as we will explore later. Indeed, despite the well-founded theories underpinning the foundations of physics, it turns out that they are not actually of great help in predicting what new physics might emerge from heading in different directions—serendipity, intuition, and clever experiments play a stronger role for Nobels.

At the time their major work was done, almost all simplifiers were supported in universities—our societies' basic curiosity was essential to underpin their explorations. On the other hand, prizewinning constructors were evenly split between companies and universities, reflecting the perceived utility of their work, and also how it is clearly primary science. As we will see later, one reason that companies support basic research is the kudos and high-tech image it gives them, as well as the recruitment of fantastic minds to the company cause. The recent demise of large company labs focused on basic science will thus inevitably reduce the contribution of industry to the Nobel Prizes, but not necessarily to the science effort in general.

In Chemistry (and biochemistry/materials), the loser has been physical chemistry (properties of molecules like how viscous they are, or how they mix), vanishing since the 1980s, while prizes for understanding how reactions occur halved. The corresponding increases have been in organic chemistry and biochemistry, which between them now account for three quarters of Chemistry Prizes. The commercial importance of efficient reactions for feedstock chemicals (starting molecules for making commodity chemicals) and for drugs has given some discoveries a high impact, while the rise of intricate molecular choreographies describing how cells do things drives more and more advances. This is reflected in the dominance of simplifiers among the biochemists. Organic chemist prizewinners are the ones much more likely to be constructors—they mostly build anew.

I will emphasize that the lack of Nobel Prizes in a field does not mean that it is not thriving and rapidly developing. It merely means that no piece of science by a distinct few individuals uniquely stands out with major impact. If we restricted our funding just to potential Nobel science areas, the web of science would be full of holes, threadbare in crucial places, with oceans of blankness. Science is not really like that.

In chemistry too, experiment dominates, with the theory-led Nobels halving to just a sixth of the total now. Emergent science does not often come from theory. Knowing how and why reactions work has not in the end been such great help in predicting efficient synthetic routes, or figuring out how the transcription of genetic information works. This balance however also reflects the disparity between the number of theoretical chemists compared to the experimentalists, who vastly overwhelm them.

Chemistry was one of the first sciences spectacularly driving industrial expansion at the end of the nineteenth century, so one would expect companies to figure strongly in its support. But more than two-thirds of the Nobels have come from universities, and less than 10 percent from companies (half as many as for the Physics Prize). It would seem that the sort of research encouraged in the major labs of chemistry and biochemistry companies is less likely to impact so widely across the entire field, than that in physics and engineering. On the other hand, in both areas the role of companies has increased over the past thirty years, as constructor science starts to make more impact.

For the Prizes in Medicine and Physiology the story is of the persistent rise of genetics, cell biology, and microbiology rising from just over half combined to more than three-quarters of all the prizes in the last thirty years. Advances in understanding human physiology (primarily diseases) have given way to the puzzle of how life works. For this reason most cell biology or genetics is simplifier science, while what constructors there are have been successful in microbiology (including immunology) and human-scale science, focused on effecting cures. Behavioral science is bypassed for now. Almost all Nobel Prize winners have been working in university hospitals or institutes (which are organized in a wide number of ways), and every single breakthrough has been experimental led. The sole theory contribution recognized was a share of the 1984 prize to Niels Jerne (a persistently country-hopping scientist) for the way our immune system can create myriad types of antibodies by combining different elements combinatorially (rather like the smorgasbord of cultures he combined). Medicine and Physiology blossomed even earlier than Chemistry, with the early breakthroughs for human health in surgery well before the recent half century of science expansion. Despite the more contemporary breakthroughs in understanding at a microscopic level, many effects on human health have yet to emerge.

DO WE USE IT?

One of the striking things that comes across from this Nobel list of breakthrough science is how many of the discoveries that became most useful, or which now generate the most money, were not done to this end, but simply to gain understanding or for intellectual interest. This is something that scientists constantly highlight when they talk to funders. They increasingly worry that attempting to choose from the start which research to support on the basis of what is (guessed) to become useful is liable to produce much less of value. But because we can't fund everyone's curiosity, funders are drawn to perceptions of utility that might be illusory.

Even a decade of weighing up by the Nobel committee supported by advisors from around the world can choose prizes in fields that subsequently run into the sand. In Physics from the 1950s, less than a sixth of prizewinners had some use in mind at the time of their Nobel breakthroughs, but nearly half these then turned out to possess valuable utility that emerged later. A typical example is the laser, which for thirty years was "a solution looking for a problem," but is now in every DVD player. Gradually more of the breakthroughs were immediately seen to be useful, after the 1980s doubling to nearly a third, with under a half so far turning out to be widely exploitable

Nobel Prizes for thirty years to:

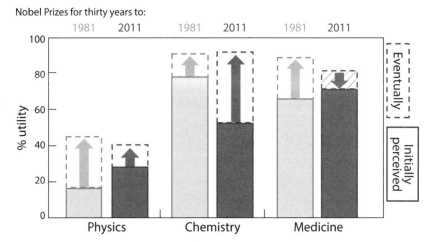

Figure 3.4: Application of Nobel Prize advances, as realized initially, or much later (arrows), for Nobel Prizes 1952–81 and 1982–2011.

(figure 3.4). In other words, earlier simplifier science has later been constructively used, while nowadays scientists are better able to see what their advances might make possible (or are forced to develop this aspect as they go along). The last half century has seen nearly half of physics and engineering breakthroughs spur constructor science, and then often new technologies. As for the "failures" that disappeared, interesting promising ideas are vital to trial as they often reemerge after decades of obscurity.

With Chemistry, the earlier epoch had much larger instant benefits—over two-thirds initially, rising to 90 percent as uses became even clearer. This was true even though more than two-thirds were simplifiers. Front-runners could almost always see where their science would lead. More recently these discoverers were much less sure (in half of the cases), but still in the end over 90 percent have led to applications. Amazingly, almost all Nobel Prize–winning chemistry is thus technologically important, almost double the exploitation of physics/ engineering breakthroughs, even though it is not always seen so at the time. This reflects the rise of biochemistries for diagnosing and treating human health, as well as the chemistry of materials and fuels that pervade our societies. Similar drives underlie the statistics for Medicine, with perceived utility in earlier times of two-thirds rising to

90 percent over time. More recent prizewinners expected over 80 percent of their work to be widely useful, but over time only 70 percent has actually delivered reality beyond the hyperbole. Recent medical discoveries take time to feed into utility, and the journey from discovery to practical application transits the passage from optimism when breakthroughs occur, to the grimmer complexity involved in trying to wrestle control over the processes of life.

To estimate "utility" here, I tried to judge if an invention has crucially allowed a significant commercial industry to emerge. Everything is useful to science, but at least one measure of its direct use to society is if someone is prepared to pay for it. Simplifier science can lead directly to utility—for instance understanding the origin of gastric ulcers directly led to appropriate treatments. On the other hand, simplifying understanding in physics normally builds up through constructor science before it is useful. Understanding that spinning electrons gyrating inside molecules could be upended by radio waves led to techniques that map locations of different molecules, and then eventually to the invention of nuclear magnetic resonance (NMR), which was refined into magnetic resonance imaging (MRI) now widely used to probe our bodies.

A brief inquiry into the highly conditional and selective Nobel Prizes emphasizes for me the expansion of high-impact constructor science. That the world's most prestigious science advances now more often than not enable widespread exploitation reflects a combination of deeper understandings, more practical motivations, changes of bias in what we esteem, and changes in what science we choose to do. These are embedded in wider shifts within the ecosystem of science. Perhaps the most critical insight though is the symbiosis between these two simplifier-constructor aspects, the yin and yang of discovery, which cyclically spur each other on in progressively oscillating partnership.

The Nobels themselves may also have to evolve sooner rather than later. It is not clear how best to reflect the role of large collaborative teams of hundreds led by more than three individuals (think of the particle accelerators at CERN, or the massive "square kilometer array" next generation radio telescope). How to deal with fields in which many small advances in many groups occur very closely in time, replacing the historic pattern in which a few groups gained all these advances themselves? How to deal with science in which the number of subfields is proliferating, so that impacts cannot be so widespread across all parts of science, but remain within the discipline's walls? The rise of information technologies, with different ways of building scientific information, or knitting science together, may make the specificity and cult of Nobels less relevant.

Perhaps the most useful aspect of the Nobels is to stress the case to fund curiosity-driven science, since many advances do not have clear visions identified at their inception. What is also clear is how science cyclically uses advances in different fields to open up new research fronts at a later time. As we shall see when we consider later how funding and careers develop within this ecosystem, the increasing competitive pressures seem to be encouraging greater focus on utility, also seen in the selection of Nobels. But to understand how science makes the journey from curiosity to utility, we have to understand how scientific knowledge is collected, sifted and appreciated, and it is to this we now turn.

4

PUBLISH OR PERISH

One of the essential components of science is writing up new results and publishing them so they form a permanent record. Society gains information, and scientists gain credit, in mutual benefit. This landscape of journal publishing shapes a tightly knit element of the ecosystem, generating a cohesive reservoir of knowledge that can be tapped. However its very openness and apparent impartiality increases the pressure of competition, creating tightly reinforcing interactions that make perceived status into a priority.

In this chapter I will mostly discuss academic scientists, because it is they who primarily gain esteem from publishing their work in the form of academic papers. In comparison industrial scientists, although they do sometimes publish papers, instead have more incentive to keep new knowledge confined within their company, and as we see in later chapters, their pressures are different.

To explore this landscape of publishing, I will first discuss how scientists find out what to read within the enormous library of knowledge they have so far created. This will make clear why the journals in which scientists publish are now ranked in a pecking order as clearly demarcated as any best-seller list. In the following section I then show how the journals themselves choose what they publish, and how they ensure it is correct. Finally I will look at what drives scientists to publish ever more papers and, critically, how they and their papers can be systematically ranked to deliver metrics that have now become crucial to their careers. The overall message of this chapter will then be how pecking orders motivate and dominate.

WHAT SCIENCE DO SCIENTISTS READ ABOUT?

For most scientists the science they read is amassed in technical articles of three to twenty pages, and collected into different journals after a critical review process. With more than twenty-five thousand peer-reviewed journals, together publishing over one million articles a year, scientists can read only the smallest fraction of them (maybe around one in ten thousand). If you thought shopping for breakfast cereal was a chore, imagine supermarket shelves that wrap right around a whole town and then trying to find a box that you prefer to eat. Even worse is that these numbers double every fifteen years (with 5 percent growth each year). Scientific patents also seem to grow a similar amount each year. How do scientists choose?

Animals find it hugely challenging to decide how to respond to information: something glimpsed might be edible, lethal, or desirable. Part of the evolution that has led to our own massive forebrains responded to this selection pressure for making the right decision quickly. A young growing brain trains neurons to detect specific flickers within visual images, such as vertical lines or dark objects getting closer. This dedicated cortical toolbox helps make rapid decisions in an information-rich world. Scientists also face a world full of rich detail. Unfortunately, evolution hasn't given useful tools for this extreme selection, so something else has emerged from the science ecosystem.

In the classic 1960s analyses I mentioned in chapter 2, Derek de Solla-Price emphasized that science output "is growing by a factor of 10 every half century, and by something like a factor of a million in the 300 years which separate us from the seventeenth-century invention of the scientific paper when the process began." Clearly at some point saturation has to set in, as Malthus warned centuries ago. Inconclusive discussions continue about whether scientists each now publish more because they "salami slice" their results into less substantial chunks, or are becoming less productive because it takes more work to produce each new chunk. However the overall increase in journals and publications is not dramatically out of line with the increasing number of scientists in the world.

Overall something near four million academic scientists (figure 2.5) produce a million peer-reviewed papers a year. This defines a level of science productivity of 0.3 fractional annual papers per scientist producing new knowledge in the published form (noting that this is not the only way that knowledge increases). Typically four authors are on each paper, though this varies enormously between fields: teams of hundreds in particle physics, solo authorship in mathematics. On this basis, saturation in total publication output will depend on the eventual stable world population of scientists. If all countries reach the 0.4 percent fraction of scientists in their population, we might expect six million papers published per year, another almost order of magnitude increase. At the current increase of 4 percent scientists per year this will be reached in fifty years, or about two scientific generations so there is still some time to adapt. Most scientists will shudder at this though because, like all of us, they swim in a sea of information and do not want to drown.

HOW TO DECIDE WHAT TO READ?

In the face of such numbers, fragmentation of attention is inevitable. A scientist working on a particular problem has to combine an intense focus on progress, while assembling some framework of what has been done before related to their problem. They may think of new methods for attacking a question, and find they need to consult research in different subfields to see how their tools can be brought to bear. They may try to see what other approaches researchers have taken recently, or they may need to look at reviews of the topic area from some time ago.

Often researchers use their problem to narrowly define what they should read and employ electronic tools to help them find previous papers, of which Google has become one impressive resource. Dedicated science literature tools such as one called Web of Science are also available and rapidly spew out thousands of papers that might be connected to a problem. The challenge all scientists must overcome to navigate this vast storehouse is setting searches that are neither too broad nor too narrow. Even then too many relevant papers are retrieved to parse.

One way to order such lists of outputs is to find out who else thought them important. Within their finished papers, scientists explicitly refer to (or "cite") previous publications of others. The resulting web of cited publications can be directly surveyed into filamentary networks of connected research. Searches of the electronic archive are then typically rank-ordered according to how many times a paper has been cited by other papers that were subsequently published. As with the phylogenetic trees that show evolutionary pathways through time, fissioning buds of interest are seen—some papers spawn many branches, while others lie dormant.

WHY DO SCIENTISTS READ PEER-REVIEWED JOURNALS?

This trove of interconnected information cannot be manually checked by a reader. Effortless and nimble navigation through it requires high levels of trust that the information published is correct, and that such maps are not flawed with ghost peaks or dry rivers. To understand how trust has developed requires returning to the birth of the scientific paper, when the idea of "peer review" was first used to provide standards of proof in the seventeenth century, during the rise of supervised science conflicts at the Royal Society in London (figure 4.1). While it has slightly evolved, the system remains in essence the same as then and forms the bedrock of the scientific ecosystem.

It works by scientists writing up their research in a suitable format for the journal they decide is most appropriate and submitting the manuscript to the journal editors. Their paper is then dispersed to several scientists who are unconnected with the research but have expertise in the topic. These "referees" write a series of comments and criticisms, which they send back to the editors, who anonymize them. Depending on the degree of agreement between these referees and their views about the paper, the editors contact the authors with good or bad news, "accept" or "reject." Editors will always feed back

Figure 4.1: Evolving discourse at the Royal Society in London. *Left*: Isaac Newton in the chair, at Crane Court off Fleet Street (wood engraving by J. Quartley, Wellcome Library archive, London). *Right*: Current summer science exhibitions for the public.

the comments on the work in an unattributed way, hence the label "anonymous peer review."

Sometimes the criticisms are so clear that the paper is rejected outright. The authors ponder the critiques (after some wailing) and decide if more research is needed to counter them, or to send the paper to a different journal for a different audience. Sometimes the criticisms are severe, but the editors offer the chance for a revised paper to be resubmitted—the referees will check if the knot of problems they identified were now solved. Despite the feelings of anger and irritation that scientists feel on reading these reviews of their work, almost inevitably the referees improve it. They are a voice from the outside, worrying about the level of proof, or the justifications for claims, or alternative ways to interpret data or theories.

While the process can be frustrating, generally the end result is much more trustworthy and reliable. Because the reviewers are not known by the authors, arguments navigate a factual plane (relatively free of emotions). This trusted space for arguments allows referees to speak truth to powerful interests (or scientists), with editors acting as mediator and judge.

This is the way the system has worked for centuries and still does perform. Worries are constantly raised about problems with peer review, but despite experimental tinkering no one has devised anything better. One repeated suggestion is to reduce the power bestowed by anonymity for referees, who can delay papers for their own competing interests with impunity, or prevent good work that might overturn their own views or reputation seeing the light of day. Publishing the names of referees with the work itself is one way to do this, since they have a hand in improving the science and effectively act as sponsors. Poor refereeing can then be noted by the community and editors and acted on. A worry though is the encouragement then for referees to accept papers (to gain publicity), while curtailing their severe comments on papers that will be published eventually. Another possible variant of the review mechanism would hide the author list from the referees so that obscuration of identity becomes symmetrical. This can reduce the influence of "big name" scientists, enhancing focus on the science itself. However this is not very practical since it is often still obvious who the authors are, when their work forms a sequence in a long research program, or their style of science and of writing are very obvious, or the equipment and approach are unique to one team only. This approach cannot be universally helpful.

DO WE NEED JOURNALS?

Another question raised is whether journals are needed at all. Why don't scientists just post their latest results onto a database or blog, and let qualified people add their comments. Indeed there are now open *archives* for scientists to post their latest papers onto (such as "arXiv and "bioRxiv"), and in some fields this really is how the community devours hot results. However the challenge is that reading a paper just posted on the archive gives little idea how it will stand the rigor of referees or not, so every scientist has to read it much more carefully. That is impossible with so many new papers—everyone is essentially refereeing the new work themselves, judging its strong and weak points. While we might differ from a journal referee's views, the duplication of effort in everyone refereeing everyone's papers makes such a system far less efficient in the forest of knowledge.

What has not taken off (yet) is a system in which scientists post their comments about papers they read. Then, I would read not only the paper, but also all the picky comments from different scientists, and use them to help me judge better the strong and weak points. As time goes on, a ranking of the validity of the paper could be devised or corrected versions agreed. Such a system is now practical and has been trialed a few times in recent years. Typical problems are the need to prioritize different comments, separating out many rather detailed, deprecating, or dull points, from subtle or insightful additions to the paper. Just who should take this role is not clear, as it is a highly demanding task. In such a system comments cannot be anonymous, because of the need to establish credibility for each contribution and the benefit from harnessing views of senior figures that you trust. But the lack of anonymity means that personal views in the critiques come much more to the fore, giving less space for the science discussions. This experiment is writ large on every contentious blog or web-accessed news article, and even a cursory trawl of the Internet does not provide solid hopes that the level of discussion remains on an intellectual plane (just look at comments on any newspaper article). Finding a blankly commented paper from a few years ago might mean what? It is wrong, or uninteresting, or written by a scarily powerful figure? Now at least such a paper would have gone through review, and, as we shall see, *where* it gets published gives some clue as to its importance. Posting to a flatly organized store of papers actually makes it *more* difficult to judge the importance of a paper, although it does democratize science and allow readers to judge. Finally, even authors of papers prefer to have a review stage before they make their work widely known, since it catches silly errors or serious problems before it damages a scientist's credibility.

THE USE OF A PECKING ORDER?

Drowning in the sea of science means that it makes sense for me (or my science funders) to pay someone who can prioritize different pieces of research—that way I can work out what to read among the huge avalanche of new science. The closer it is to my detailed expertise the more I want to read these details, but the further it is the more I want others who I trust to provide a distilled view to me. This is one reason for the emergence of a pecking order among the journals—it helps enormously in deciding what to read. We could let posted comments decide on "importance," but this yields fickle movements of opinion. The peer-review system has proved extremely effective in giving all new research a relatively fair hearing but at the same time allows some discussion about the importance of the work and estimates of its likely influence.

What has emerged is a journal habitat, in which there is a fairly clear ranking of importance. Each scientist wants to get their paper published in as significant a journal as possible. Even though their paper is accessible wherever it sits, the higher the journal ranking then the more likely it will be read, the higher its likely final impact, and the higher the esteem the authors will accrue. Faced with choosing to hire one of two scientists with the same number of papers published, but with one publishing mostly in higher-ranked journals, then this will be taken into account in recruitment.

While many distortions skew this comparison, if the two scientists come from the same subfield it can be valid. Other factors might enter into our choice though. One might be their temperament (since if one scientist is more humble they might never submit their really excellent work, or make aggrandizing claims for it); another might be their energy (since fighting to get a paper into a higher-ranking journal takes more time, leaving less for the science); others are their luck (though one would hope this averages out over a sufficient time so is hardest for younger researchers who haven't tried much yet), their ambitions (since choosing hard problems that you can make only slight but useful progress on may not lead to high-ranked journals), their boss's focus (since they may have been directed to problems less trendy or leading to lower-ranked journal papers), their way with words (since higher-ranked journal papers demand a higher standard of presentation and English), or their charismatic vision (since reporting the science in disengaged facts may not compete against enthusiastic articulation from competing papers). However, since all these aspects might well be relevant for a hiring selection between the two scientists, such measures still have some meaning.

As with any pecking order it has to be sustained through competition over a finite resource, and for journals this is provided by a limit on the number of papers accepted each week (or journal issue). Paradoxically such limits are now artificial since print versions of journals, which are limited by deadweight, spine binding, and the tangibly physical nature of the medium, are becoming irrelevant. The self-imposed cap on the rate of published output from each journal gives two things. The first is utility for the reader, since it provides measures they value, including selecting only the few "best" papers to read from a larger competitive pool. The second is prestige for the journal, since the perception of its ranking depends on how difficult and desirable it is to get published in.

Perhaps it is clear that these are self-reinforcing aspects. A journal with high prestige strongly limits the number of papers it publishes, driving competition to get in to high levels, thus increasing its prestige and its desirability. Predictably then submissions to such journals have increased faster than the overall increase in science outputs. This poses significant problems to journal editors, caught between accepting more of the very highest quality output, or retaining their page limits. If a journal turns away too much of the best work, there is an incentive for a rival journal to take it on, expanding into their prestige or their target market. Journals, it is clear, operate like any other competing species, or brands in the commercial world, with stratification, niches, customer base, and branding. It is this signposting through brands that helps scientists decide what to read.

Journals with a winning formula for high impact are spawning ever more descendants: *Nature* has produced *Nature Physics*, *Nature Materials*, *Nature Nanotechnology*, and a host of others (figure 4.2); *Advanced Materials* has produced *Advanced Functional Materials* and others; *PLOS ONE* has produced *PLOS Biology*; and so on. The combination of an existing income stream, editorial team, community,

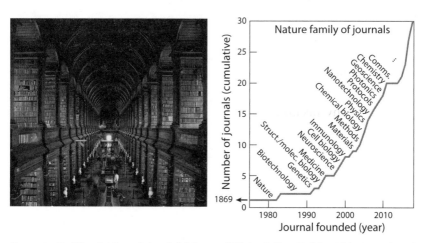

Figure 4.2: Proliferating knowledge. *Left*: Library of Trinity College Dublin. *Right*: Spawning of new journals from the original journal *Nature* founded in 1869.

and publishers makes it much easier to spawn a new journal than any outside team could. Naturally there is mother-daughter competition for resources and limelight, just as within mammalian families.

JOURNAL DIVERSIFICATION

The other reason to split apart published science into many organized journals is to help readers find what interests them. In the days of printed journals, I might browse through an issue skimming articles that seemed interesting, and then focusing on a few papers reporting critical advances for my own efforts. To make this effective, articles had to be grouped by subject area, and this has persisted even in the digital age because we know what to expect from each journal when it has a brand.

Subfields now have topical journals. Typically their shared language and aspiration allows authors to briskly state the details of their science advance. Each subfield might have its own pecking order of journals, with a top one that is most competitive and exclusive. Then each wider discipline has a more broad-based and even higher ranking journal in which the most widely relevant papers of the subfields are published, of potential interest not just to the scientists in one particular subfield. Different journals in each discipline might be broad but still have a focus or theme, for instance on science related to solving a wider problem rather than the subfields that are contributing.

Above this are journals that purport to reach across many disciplines, culminating in a few premier journals that stand for the pinnacle of all science. Papers published in such journals are supposed to be of interest for every scientist. We will see that the selection of papers they identify is crucial in what all of us across society hear about current science. But equally, this partitioning into journals helps scientists themselves decide what to read, and how to evaluate the claims that are made. So journal hierarchies are here to stay, but how do they form?

WHICH JOURNALS ARE IMPORTANT?

Scientists are bred to enjoy avoiding consensus. Details matter, differences are amplified, agreement more quickly passed over. However, for the journal ecosystem to work, there needs to be an accepted definition of the pecking order, allowing for slower changes in ranking over time. Competition in the ecosystem thus leads to the need for agreement on this pecking order. No longer is a gut feeling enough, but numerical measures of importance need to be devised.

IMPACT FACTORS

The clue to how such numerical weighing works is in the citations to previously published work embedded in a scientist's written-up papers. This vast network of citation links can be collapsed into numbers, the simplest of which is a single value. The number of citations to a paper by subsequent articles is a measure of its "impact" in science. If my paper has been cited many times, while yours only once, then more people are interested in my work, or need its results, or feel it justifies their own scientific interests, or even that it is wrong but in tantalizing or provocative ways.

Just as weight is no guide to a person's importance, citations are a very coarse measure. They depend highly on the culture of a subfield. If in my field we always include a hundred citations to previous work, but if in another field they put in only the most important one or two, then citations will always be lower in this other field even though a paper may have equal impact to mine. Or if in another subfield, they publish only one long and detailed article a year while in mine we publish many small succinct articles (called "letters"), then the distribution of citation numbers will end up very different. Also citations might not show me the influence of a paper on the many more people who read it but don't leave a permanent glowing trail, imprinted via a new citation to it.

The simplest way used to create a journal ranking or "impact factor" is to find the average number of citations received by all the papers it published over the last few years (normally the two previous years, see figure 4.3). For high-ranking journals, we expect everything they publish is important and every paper will get many citations very quickly. In the highest-impact broad science journal, *Nature* (which has reigned supreme for several decades), this average is currently around thirty-eight (though the *New England Journal of Medicine* has an impact of sixty). On the other hand, the distribution of citations from articles in topical journals in many subfields is very broad. The most likely number of citations is none, some have only a sole citation, far fewer have

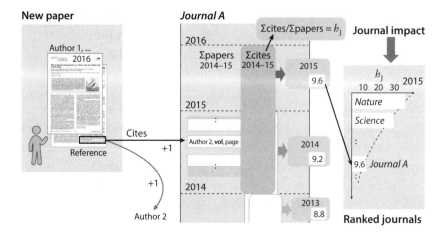

Figure 4.3: When a new paper is published by author1 (*top left*), each reference it makes to another paper in *journalA* adds one to the citations in *journalA*'s lists. Summing all the citations to papers within two years of their publication in *journalA*, and dividing by the total number of papers published in the same period, gives *journalA*'s impact factor h_J (*middle*). In each field the ranking of journal impact drops rapidly (*right*).

two, and only a tiny number have much larger numbers of citations. It is definitely true there are important papers in obscurer journals whose potential importance was originally missed by referees or the authors, or papers that held a critical detail needed for later research. But these are the exception, not the rule.

Listing all journals in order from the highest to lowest, their impact factors drop rapidly (figure 4.3, right side). How rapidly this drops off depends on the discipline, from mathematics (a less severe pecking order) to chemistry (intensely competitive). Such behavior is also common for hierarchies in naturally emerging ecosystems. For any randomly selected journal in biology the average impact factor has been 2 in recent years, in physics or chemistry it is 1.25, while in engineering or social science it is 0.5. So an average paper in an average journal will be cited only once in its first two years after publication.

SELF-FULFILLING PROPHECIES

Part of the problem with these journal impact factors is that they tend not to change over time, so the pecking order becomes ossified. Scientists want to publish in the highest-impact journals as then their work will be read not only by people in their subfield but also by a much wider audience. If they succeed in the face of stiff competition, then not only can their peers judge the research to be strong, but so can their departmental colleagues in different fields, the administrators of their institutions, their funders, and all of us too. So publishing a paper in a high-impact journal is significant for their career prospects, and to their ability to be funded to do more science.

This desire for quantifiable approval locks up the science ecosystem by blocking changes in journal impact. Scientists are much more likely to cite papers in high-impact journals, not just because they are better, but because they are more visible. If in my paper I want to cite previous work in the field that supports how important my research is, then I will have many possible choices from the published archives. To help me evidence the buzz in my field, I select papers recently published in the highest-impact factor journals. Since I boosted its citations, now others will too, so the escalation continues. And as this boosts the journal's impact factor, its editors are also content to believe they selected a good original paper to begin with. Unfortunately, this added esteem is a self-reinforcing part of the system.

FILTERING THE WHEAT FROM THE CHAFF

What maintains journal publishing to sufficient satisfaction for all scientists and funders is the peer-review bedrock, so that only really good and really correct science gets published in the stronger journals. But so far this does not tell us how different good and correct pieces of science are elevated to the highest echelons in the pantheon of journals. Here another component of the peer-review process combines much more subtle viewpoints of the editors and referees. Since the highest-impact journals are so competitive, only a fraction of the papers submitted can be published in each. *Nature* receives ten thousand submissions each year and publishes fewer than seven hundred of them. Hence a first sifting is done by the editorial team (normally with expert advisers in different areas) who judge whether this new research not only is timely but also might have sufficient impact to compare well to the level published typically in the journal.

This judgment attempts to predict the importance of science, something that legions of pontificators, futurists, and prophets have attempted poorly for generations. To help them, authors are specifically asked to justify the potential impact of their research, asking them to hold hands in this leap from science to speculation. The authors of the research have to weave a vision full of intrigue and luxurious blossoms, fruiting a promised land enabled by their single step forward. The editors are looking for something else: they seek papers that will be the most widely talked about as possible, whether in a cascade of citations throughout the web of knowledge, or flooding across the media of the globe. Sometimes these are very different goals, but at other times they knit into a tight-meshed gauntlet flung down in front of the community.

Important and excellent new science can find itself excluded by this editorial filtering just because it does not lend itself to portrayal in any straightforward way to an audience beyond science's niche. Dissemination favors those who find the right way to weave tantalizing messages into their work, who adopt metaphors that blend some measure of science fiction into the intuition behind their results, or in some cases who suppress clarity for showmanship, hyperbole, or complexity. Well-produced visuals showcase both vision and science, conveying them to people outside the direct field. But these creations are extremely time intensive, fed by researchers' time and funders' money, readily produced only in well-funded and successful labs. Within large high-capital transnational collaborations (such as particle physics), even more such deployable resources are available, driving a bias in impact between subfields.

If papers get through this editorial filter, they are sent to referees with experience in high-level publishing (figure 4.4), already conditioned to defend the journal's prestige. Referees are asked not just about the science, but about its suitability for this journal and its readers. They have a vested interest in protecting the journal's reputation since their own previous work would be devalued if the floodgates opened. Such a system is innately conservative—authors have to work hard to convince referees of startling advances beyond the mainstream view. Counterbalancing this are editors wanting to ensure their journal picks up the highest possible impact factor, not

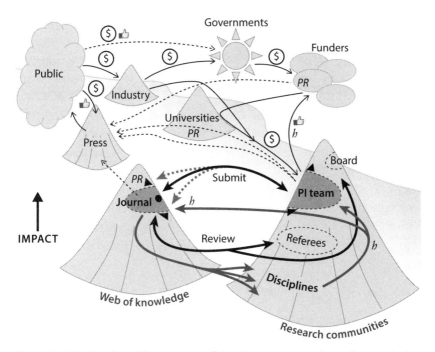

Figure 4.4: Mapping the publications part of the science ecosystem shows the interplay between researchers (led by a principal investigator or "PI") and journals (together forming our web of knowledge). PIs submit their manuscript to a journal, who ask trusted referees from the community to evaluate it (solid arrows). Eventual publication to this full community leads to citations, which increase esteem of both author and journal (gray arrows). Funding streams ($) are influenced by resulting esteem (increasing citation h, and plaudits, i.e., thumbs-up), eventually reaching the press and public through press releases (PR).

missing out on any profound new science result to a competitor journal. The result is a subtle conflict between authors, referees, and editors about what is significant and new.

In hindsight, very little in science is really *really* new. Gratifyingly, science intricately hangs together so results in one area corroborate results in another. Large gaps or problems tend to appear in several places at once and are rarely neglected. Mostly, new science comes down to things we didn't do before, either because we didn't think to, or because our funds and technology are limited. Possibly it emerges from a small detail interesting only to a subfield, or because a number of advances in different disciplines enable new research that depends on bringing them together. Such advances are infrequent, and most science is more predictable, though fascinating. Despite disagreements on what is "surprising," more than a third of papers in the journals *Nature* and *Science* are afterward cited as the top 1 percent of papers in that year and field. There is thus much consensus on importance.

Arbitrating from several independent referee reports, editors rule on the correctness of the science and the appropriateness of their journal, assessing carefully the potential for citation impact. Authors might receive an e-mail saying that the work is liked by the referees, but not interesting enough for a general audience. Emerging fields that don't yet have an audience can fail this test— indeed they don't have many readers yet despite possible import. "Importance" blends both subjective and objective viewpoints in contentious conflict.

WHO REFEREES?

The gatekeepers to journals are the referees of the submitted manuscripts, who are selected in several ways. Authors are asked to suggest a few people who can referee and of course select those they feel are highly respected but also supportive of their work. The journals select extra referees from among papers that the authors cite in their manuscript, as well as scientists in the general field who have published before in their journal. There is no formal training for becoming a referee except learning on the job, seeing the reports we receive on our own papers, and perhaps a little guidance from a senior colleague we consult. For this reason, different subcultures grow up around refereeing, guided loosely by the editorial teams, but rather conservative to change. The style of reviews has changed little for a century.

Referees read papers in varying amounts of detail. Some clearly skim through and form a rapid judgment about whether the science is really novel, providing a brief paragraph. Others read every sentence carefully writing a line-by-line critique of ideas, methods, analysis, lack of references to previous works, and unwarranted conclusions. All of them are enhancing the science reported considerably, and expending considerable amounts of their time and skills. Perhaps it is then surprising that they are never paid for this.

THE ECONOMICS OF REVIEWING

The average time that a referee spends on a paper varies greatly, and little hard data is available, but publishers estimate that four hours work generally provides a comprehensive and useful review. Young scientists without a lengthy background in a specific area might have to spend more time reading the relevant literature, but all scientists find themselves forced to read cited papers from the manuscript that are unfamiliar. Scientists review anything between five and fifty papers a year, perhaps twenty on average for an established scientist. So overall they are contributing a hundred hours each year, something like 5 percent of their paid working time.

Since the reviewer does not get paid, their employer is funding this publishing activity. Why would they do it? The peer-review system is so crucial to the workings of science that funders, whether civil servants, universities, companies, or institutes, see reviewing as an essential part of being a scientist. Unlike other parts of the science ecosystem, this activity has not been made competitive (since reviewing remains anonymous) and is seen as a system-wide "good." In fact only recently, as universities become more like businesses in some countries (particularly the UK and the United States), has there been any attention to the amount of time scientist devote to such public contributions.

While society benefits as a whole from the increased reliability and truth embedded in science, it is the publishers who most directly financially benefit. Their costs do not include this professional consultation and are restricted to running the refereeing and editorial processes, publishing (digitally or physically) the journal, and marketing and distributing it. By and large, authors themselves provide the copyediting and typeset the text themselves. Sporadic revolts by scientists have boiled up over the past twenty years, arguing that their own contributions as authors and referees financially reward only the publishers. Strong arguments declaim that publicly funded research should be available to the funders, that is, the whole of society, not only to those who can afford to pay for the technical journals. This has become much more of an issue as science becomes of greater interest to a wider cross-section of our societies, who want to access it electronically but who find that they have to pay the publishers to read an article. Typically this might be $50 per article, and because each article has to be unique there is no possibility of competition bringing down the price. Many countries are grappling with this issue, and there have been moves to insist on public archiving of all publicly funded research. More recently owing to the elimination of large-capital infrastructure from the digital publishing process, a host of new journals have been set up that are free to access ("Open Access"). Here the authors now pay to have their work published instead of readers to access it, while some journals allow both financial models.

Publishers argue that the organization of the journals is a valuable function that we need, and that the reason for them being in business is to make a profit. However the business of publishing journals is extremely good, with stable margins of more than 30 percent, more than any other part of the entire publishing domain. This golden goose is also signaled by the lack of journal titles ever being closed down and instead the perpetual increase in the number of journals published. Many scientists see publishers as parasitizing a publicly funded resource. To make things more complex, some publishers are not commercial concerns, but charitable organizations run by the same disciplinary institutes and societies that hold the vast science meetings we will talk about later. Profits they make from publishing are partly used to support other (less visible) aspects for the good of the discipline as a sort of tax on research outputs.

Despite their more egalitarian cost model, new Open Access journals are not widely displacing established journals, very visibly demonstrating the tangle of feedbacks festooning the science ecosystem. Scientists have strong motivation to publish in the highest-impact journals. Publishers have strong interest to keep the same journals at highest impact, and preserve the pecking order. Funders compete to show they have delivered the most important science. Perhaps the only tears reliably shed are by universities, institutes, and companies that have to find budgets to provide their scientists' access. At $4,000 per year for a typical chemistry journal, it is no wonder that 65 percent of the budget for academic libraries is swallowed by journal subscriptions. Librarians are not primary actors in the publication ecosystem and have to lobby for new funds through increasing student fees or administrative costs (research budgets rarely cover full publishing costs). Open Access exemplifies a paradox of science I will return to, with researchers being charged to give away their work.

DATA FOR ALL?

Another facet of the Open Access debate is whether all results and data should also be provided for free. For publicly funded research why cannot everyone get to see all the raw results, and not just the parts selected and interpreted by the scientists heading the teams. Just as for publishing online without peer review, this complaint forgets the time-saving value of selection and summary. Scientists already have a huge choice in how and where to publish their results. Sometimes they select a few choice aspects and create a focus that will appeal to a high-impact journal and its audience. At other times they might publish a long, overarching study that will form a trove for generations of future researchers. Picking over such data driftwood is a livelihood for many and strongly enhances the value of the original science investments.

One of the needs for any publication in science is to tell a story because rarely is the primary data itself of much use. There are too many details about exactly how the data was taken, what experiments or theories were selected for focus, what aspects are artifacts to be discarded, and what the appropriate way to analyze the data is. Referees argue with authors about these when the paper is written, but mostly they have to defer to the original scientists' decisions after checking that the scheme as a whole seems robust (by asking awkward questions). That this works is due to the way each scientist builds up their own brand identity. In any subfield, scientists and research groups acquire a reputation for flaky results or excellence and robust output. On first reading your work, I won't know if you saw your results one time out of many tries, and if you are showing me only the parts that conform to your arguments (a particular problem in medical research). But after a number of my colleagues tell me they tried the same techniques without success, I will be a little cooler in my enthusiasm and a lot more questioning of you. Such opinions supplement published knowledge but are notoriously hard to pick up for young scientists, who are shocked to be told that not everything published is correct.

A really experienced scientist knows what smells right, and how to hound surprising details until they deliver a gobbet of insight. Sometimes this reveals the flaw in a carefully planned experiment or study. Sometimes what goes wrong is much more interesting than what has gone right. But just archiving publicly all the raw results without a story would yield little unless massively more information is provided. Either the team has to remain available to be questioned on all these details, or somehow the details have to be codified and added to the data archive too. In either case, it would make publication much more expensive in time and resources, with diminishing returns. Forcing scientists to do this without providing resources just decreases the amount of science possible. Even in my own research, where we aspire to capture all this implicit knowledge automatically along with the data, it is still an extremely difficult proposition. How should scientists capture all the discussions in the development of their research project, or trace back the calibration of their detectors? Better value might be to capture all calculations made on the primary data that give formal outputs such as graphs. Perhaps the most surprising thing here is that to make the scientific ecosystem work, it is important to leave things out when reporting work. These might be data that cause confusions, excessive details, or things that didn't work. What gets a paper published is communicating a story, not building an archive of data.

WHAT HAPPENS TO REJECTED PAPERS?

Of course papers that are rejected are not abandoned. Normally the reviewers unearth important issues and an improved manuscript emerges from rewriting and is submitted to another lower-impact journal, more specialized in the subfield. Hence papers are filtered down through a cascade of journal interactions, until they find a suitable level. Papers with major problems are generally ejected from the whole process repeatedly at every stage. Often different journals end up using the same referees because they are the most appropriate for the subject, so continually encountering again the same implacable objections. Papers that are poor science trickle down to the lower echelons of journals, where some make it past peer review even though they ought to be left to die in isolation. While experienced scientists automatically interpret the uneven terrain of the science literature, young researchers or nonscientists often point to such articles as peer reviewed and thus deemed correct. While the pecking order helps stabilize confidence and belief in the scientific world, it is less transparent or apparent from outside.

A perverse aspect of the self-differentiation of journals and their pecking order is that each demands a different format and style, with differences in all details down to their font type. Different scientific disciplines stress distinctive aspects of the research and favor different presentations—health statistics are treated very differently to mathematical derivations. For each paper rejected from a higher journal, the authors have to pick the next journal to target and recast their work. The incompatible formats make it harder when devising electronic tools to digitally connect the network of science knowledge—something that new AI tools must overcome. A publishing format (comparable to hypertext markup) in which all papers are written would be incredibly useful. Reformatting could then happen on the fly for readers, at the cost of giving up some design control. But because this would reduce the differentiation of journals, they would fight against any such move.

Perhaps most amazing in the process is how fair referees generally are. The current anonymous review process seems to provide a helpful mind frame that encourages respect for the scientists who did the reported research, and to focus on the robustness of the results. Occasional problems come from inattention, conservatism, or maliciousness, but while the methodological aspects of papers may be debated back and forth, almost always the results are better. An interesting experiment by the editors of *Economic Inquiry* highlights support for peer review: they forced referees to respond with only a "publish" or "reject" answer and allowed authors to ignore suggested revisions (since apparently these can make economics papers *more* opaque). Still, all the accepted authors revised their papers using referees' comments. Having critics devote time to your science, able to stand back and give practical advice, provides an amazingly useful service. Even if not embedded into the publishing process, we would still want it somehow in our ecosystem.

WHY DO THEY DO IT?

Scientists are brought up with peer review, and the ecosystem would be almost unthinkable without it. But the time and effort devoted to it means that while it is a public good, it is also a drag on their own research. In such cases, the public good is often eroded by free riders ("hawks") in the system. In fact, however, reviewing is a very refreshing and efficient way of being exposed to new science and accessing more randomly selected knowledge, and it is often highly stimulating. There are indeed papers where a referee feels as if they are writing a lecture to the authors about what to read from the past or how to interpret their results properly. In this case, their efforts are more like unrecognized coauthorship, and journals are full of acknowledgments to helpful anonymous referees. But at the very best it is a pleasure to suggest to imaginative authors something that helps their work onward. Seeing good science is rewarding in its own right.

Besides these reasons there are some very personal ones too. Refereeing allows me to keep track of my competitors, to see which direction they have been going in and to get insights from their slightly rawer presentations as well as to be able to argue with them on detailed points. It also allows a reviewer to demand the authors cite their own previous work, at the cost of partly losing anonymity. This intellectual ransom goes, "if you add a citation to my work, I will be more likely to agree to your work being published," so the advice is often followed, despite the sometimes inappropriate new linkage. This is balanced by the benefit of encouraging citations between subfields, reconnecting where common themes have been ignored.

Refereeing has remained one of the few theaters where arguments can be fierce and detailed. For this reason, it is crucial that referees are not recognized by authors, even when subfields are small. Referees can disguise their responses, changing their language style or grammatical accuracy (since so many contributors are not native English speakers), and emphasizing citations to other groups' work besides their own. Referees sometimes reveal themselves to authors after publication, but this is not helpful for the system as it introduces an obligation to repay the debt in reverse (which is why it is done, fueling tribal networks).

While scientists often complain about delays introduced by the peer-review system, this attitude is mostly based on career or competitive personal issues. From the viewpoint of building a reliable web of science knowledge, delays of months are a very small price to pay for the improved reliability of the knowledge added. For the scientist though it can mean the difference between being first or second to publish (which is everything for personal awards), or affect career decisions about the next research appointment, or funding decisions. So despite journals competing to provide the fastest possible conversion of manuscript into publication, this is not of clear advantage to the science community and leads to a steady escalation in the jostling and competitiveness.

WHY DO SCIENTISTS PUBLISH EVER MORE?

Many of these discussions around publication contaminate what scientists have felt to be the most attractive parts of research—seeking and extending knowledge, and the addictive pleasure of new insights. So why do they try and publish more and more?

CLAMOR FOR ATTENTION

Competition more than anything has increased the clamor for recognition among scientists. Just as one of the emerging genres in media has been the elevation of ordinary competitors to stardom in TV reality shows, some scientists seek importance and fame. This circus is rather distinct from the detached view of a web of knowledge steadily added to by questing tendrils of research around the world, which has little care for who, where, and when. Successful science is often peremptorily cleaved from its inventors to become ubiquitous in hundreds of research labs. Once no longer controlled by any individual, ideas can be more urgently driven and spread by many supporters who co-own it. In other cases ideas may wither unless a determined individual relentlessly champions how it can unearth new science possibilities or foster a different approach. For their efforts, such scientists can become unbreakably associated with a technique or advance, even despite their best efforts to detach themselves (and their brand) from it.

Scientists need several feedstocks to thrive: colleagues, infrastructure, funding, and a community of peers providing motivation. The less a scientist feels noticed, the more they can feel frustrated. Unfortunately rarely are scientists given feedback on their work, apart from critical reviews. Colleagues in my department come to tell me more about the grant funding they won, than about their new and interesting science results. Previously a scientist embedded in a topical field would get some idea of who read their work from colleagues coming to talk to them about it, but with the rising number of papers and scientists it feels like informal recognition is becoming eroded apart from a favored few who are elevated to stardom.

In response, scientists have started to become more and more concerned about the number of people who cite their papers, coinciding with the digital enablement in calculating simplified metrics. Just as journals have impact factors devised for them, an impact factor can be produced for each scientist by looking at the citations to all their papers. Arguments about which is the most suitable measure are currently waged in the fashionably growing disciplines of "infometrics" and "scientometrics." The most common one is the h-factor, popularized by Jorge Hirsch in 2005. Take a rank-ordered list of an author's papers from the most- to least-cited, and record the position at which the citation number drops below the position number in the list (figure 4.5, *lower left*). So an author with h-10 has ten papers all with at least ten citations. This metric rewards consistency in high citations rather than rare mega-successes, rewards experience over young stars, rewards wide collaborators rather than lone specialists, rewards fields that cite more, and rewards those who stay always at the trendy edge of all research.

Citations are actually a form of money, most similar to the idea of bitcoin. Doing research generates publications that are cited, increasing their value to their owners (in this case their authors, as well as institutions and funders). The verifiable work that went into the creation of the citations and their robustly checked authorship creates

New paper

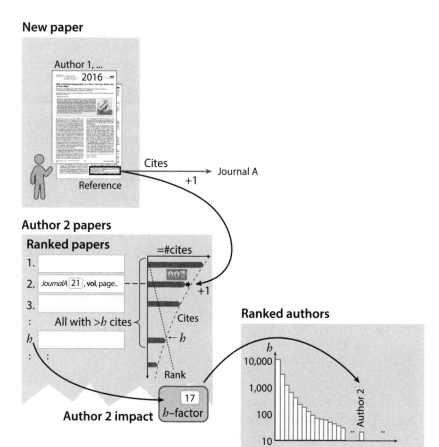

Figure 4.5: When a new paper is published by author1 (*top left*), each reference it makes to another paper by author2 adds one to the citations in author2's list. Ranking author2's highest-cited papers gives their *h*-factor as the point where all higher-ranked papers have more than *h* cites (*lower left*). Only few authors have large *h*-factors (*bottom*).

a type of currency, which is partly exchangeable for other types of money (such as salary or bonuses).

The desire to characterize a scientist by a single number resembles similar efforts that characterize baseball, cricket, or basketball players. It hides a multitude of individual approaches and skills and confounds the very details that really lead to specific success. However it is tempting for those in difficult positions, such as trying to judge hundreds of candidates for permanent jobs or promotions, and particularly in sorting among a large number of people. Citations and metrics are becoming one component that *all* people who judge scientists now look at, although hopefully they also use other information to form judgments. As a result, citations and notoriety are now very important for enhancing a career. This is most extreme in China, where publishing in a high-rank journal gains a financial bonus and is the only way to access a permanent position, creating massive pressures (and incentives for fraud).

Because publications are alleged to indicate productivity, national governments and funders interpret citations and output to compare the effectiveness of different institutions. Administrators in some universities issue edicts about the number of papers to publish in high-impact journals each year. This leads to increases in competition, and an inflation of the purported merit of many articles.

LETTING THE GINI OUT OF THE BOTTLE

Just how differently cited are different scientists? One measure is to take the total citations or *h*-factors of each scientist, and then compare their distribution (figure 4.5, *bottom*). So we count how many researchers have one journal citation, how many have two, three, and so on up to the maximum. The author who has been most cited in recent decades is Bert Vogelstein of Johns Hopkins in the United States, who with over five hundred papers has garnered citations from 293,153 papers of other people since 1976 (and is still active). Of the top fifty most-cited researchers over the same time, around half are in genetics and molecular biology, and all are in the biological or medical sciences, reflecting both the rise in these areas and their tendency to cite others more profusely. On the other hand, most papers receive only one citation, if any. It is increasingly unlikely to get more citations, and really high-impact papers are rare.

One way to survey the range in output of different science researchers is to look at their Gini index. Devised almost a hundred years before the *h*-factor, the Gini index is another single number comparison, this one aiming to describe the inequality in any

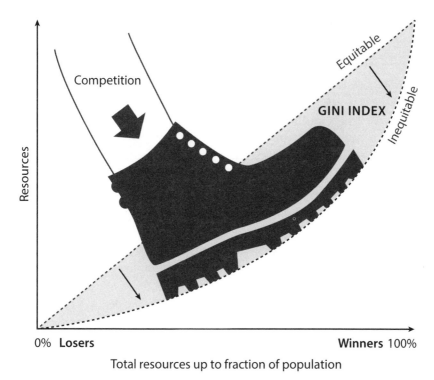

0% **Losers**   **Winners** 100%

Total resources up to fraction of population

Figure 4.6: Ranking the resources of each scientist, and summing them up to any particular fraction, would produce a straight line (dashed) if all the shares were equal. The fractional area (shaded) between this equality and the real distribution (with more to the winners, curved line) gives the Gini index.

distribution. A completely equitable distribution (with zero Gini index) would have 40 percent of people having 40 percent of resources, 70 percent with 70 percent resources, and so on. Real distributions never look like this, so for instance in ecology 20 percent of the scientists have more than 82 percent of all the citations. The Gini index sums up these fractional differences to give a number between 0 (equitable) and 1 (winner takes all) (see figure 4.6). For ecology, the estimated Gini index turns out to be 0.8, which is very unequal. Similar ideas are used to measure biodiversity, where the cumulative proportion of species is plotted against cumulative proportion of individuals. It thus reflects how diversely spread are successful individuals or species in each system. Rising Gini factors suggest something amiss with any system, from incomes in our society to mammals on our planet. Countries with high Gini coefficents for income have fractious populations, and often high levels of violence.

WHOSE DISCIPLINE IS MORE IMPORTANT?

If I am a molecular biologist, I am nearly five times more likely to get my breakthrough paper accepted into the journal *Nature* than if I am a physical scientist. And the most common fields to be published in *Nature* are all biological (in order: biochemistry/molecular biology, which is involved in 42 percent of papers, followed by genetics/heredity, cell biology, and zoology), while engineering figures in only 7 percent of breakthrough papers. Why is this?

If the editorial team at *Nature* is biased, then they are not alone across the scientific community—the frequencies are almost identical for the journal *Science* (over the past five years). The most likely explanation is that findings considered of high impact and interesting for those outside the field must overturn accepted wisdom, or be "surprising." In hindsight very little is really surprising (since ideas have to fit previously accepted facts as well as new observations, slotting into the existing web), but it is easier in biology where there is still "wiggle room" for new paradigms. In engineering, the network of knowledge is dense and robust, so that new surprising facts are mostly localized to connecting regions between established areas where deformations of the knowledge network are a bit easier.

The distributions of papers have been remarkably stable over time. Over the past twenty years, each year *Nature* published 40 percent biological/molecular biology papers, 40 percent genetics/heredity, and 30 percent cell biology (there are some overlaps), while physics languished at 10 percent. There have been some changes though. Both plant science and zoology saw steep rises in the late 1990s followed more recently by falls, over which period the use of plant models (arabidopsis—a small flowering plant related to cabbage and mustard) and animal (mostly mouse) models for breakthroughs in genomics and proteomics burgeoned and saturated. Chemistry has surged, doubling to 8 percent more recently as it unlocks new paths in pharmacology and nanoscience. The other significant new focus has been environmental and ecological sciences, which at the turn of the millennium tripled to 10 percent, reflecting the intensity of interest around quantitative evaluations of planetary change. The losers have been biophysics, immunology, neurosciences, and astronomy, areas that promised hugely in the past and expanded, but are not yet managing to deliver on this promise for the high-impact journals. The typical traces of fashions and fads can be seen in these strata of the publication record.

We can also see how different disciplines interact with each other using citation mapping, which tracks links between many papers in different subfields (figure 4.7). The links between molecular biology and medicine are particularly highly interconnected, while citations between physics and chemistry form another strong bridge. More detailed maps allow the compartmentalization of science to be studied, and how its interactions are evolving.

WHOSE NAME ON THE PAPER?

As results are written up for publication, one senior author takes a lead role, normally the person who motivated or headed the research program. By contrast, many people often contribute to the full effort—in particle physics this now reaches up to three thousand authors, a substantial fraction of the entire field. In other areas a small advance might be attributed to a few authors, but the entire subfield has shaped its emergence, the combination of competitors and collaborators as a whole. How then does one decide who should be included or not as a formal author of a paper, adding to their h-factor? This issue causes much emotional ache and many a rift between scientists, fracturing their combined wisdom forever.

The acknowledged recommendation is to include all those on a paper who have contributed enough to be able to argue its contents with other scientists. However there are occasions where the spread of disciplines involved is too broad to expect everyone to have expertise on all parts of a paper. Another model includes only the essential contributors, but this is a gray area—ideas may be sparked by discussions with competitors at a meeting, or with referees on a previous paper, or colleagues from several years ago. Or it may have

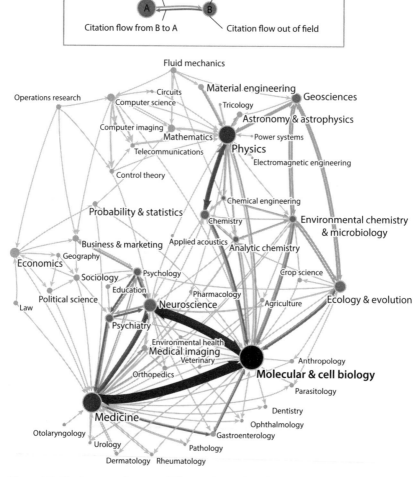

Figure 4.7: Citation maps between different science disciplines. Reprinted by permission © (2008) National Academy of Sciences, U.S.A.

been stimulated by a research student who finished their work a year before and has since moved elsewhere, or have emerged from a group discussion between many researchers, but then realized by only one of them.

Instead of a clearly definitive roll of authors in black and white, publications should really associate a long list of names written in slowly graying-out type with shrinking fonts. This may indeed become more practical, but already some journals require scientists to outline who did which part of the work. While dividing attributions carefully can be important for the people involved and their careers, the science network of knowledge itself is helped not a whit by such arguments. Giving equal weight to all contributors strongly encourages collaboration. The down side is the increasing number of authors on each paper, the slow devaluation of esteem associated with publication, and the relentless focus on impact.

HOW MANY PAPERS ARE READ?

Scientists want their work to be read. Mostly, papers will be cited only once within two years. A decade after publication, an average number of citations is ten to fifteen in physical sciences and twenty to thirty in biological sciences. The most cited paper of all time (with 329,857 as of 2017) is on a scientific method (methods are often more cited) by Oliver Lowry, measuring the protein concentration in a sample using a color change from yellow to blue. He is now almost unheard of apart from this. On the other hand, almost half of papers in some disciplines are never cited at all.

Detailed studies in recent years show the fate of a paper has a tipping point. Most papers sit in a distribution where if they have slightly more impact, they are slightly more noticed. However above a characteristic number of citations, papers are much more likely to be cited than this random obscurity effect (figure 4.8). The tipping point depends typically on each subfield (and is typically twenty to thirty). Such effects imply feedback: citations to an academic article are proportional to the number of citations that the article already has, something de Solla Price termed "cumulative advantage." Amazingly, without knowing any more details statistical models capture these behaviors (in itself a typical simplifier science outcome). It seems that as a paper becomes visible enough to a community, it becomes shorthand for referring to the subfield or a technique, rather than for any details. We will later see how journalistic reporting of science similarly depends on constellations of memes. Here such "preferential attachment" has a side effect: the citation rich get citation richer (even when they are wrong).

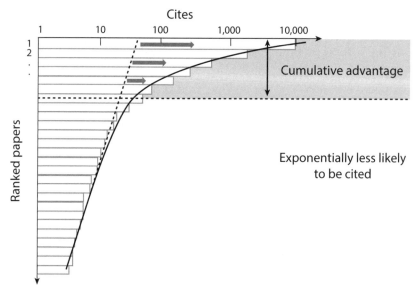

Figure 4.8: Ranking science papers by their citation count shows that a few that are highly visible (shaded) attract far more citations than expected. This effect arises only for the top ~20 papers.

Since journals are now accessed through the Internet, they can count the number of times that papers are accessed, suggesting the idea of a "readership factor." With many confounding factors, such as search engines passing through, or accidental downloads of topical sounding titles, it still appears there is a relationship between citations and readings—estimates suggest between fifty and a thousand times more readers than citations is typical, but the relationship varies widely across fields.

What about those papers that are never cited though? Could we not spot them beforehand, and save resource for more "useful" science? Such efforts might be misguided though. Detailed but arcane papers can forestall a later researcher from repeating the work themselves, saving an enormous amount of time and resources. And yet the paper still might not be cited, for instance if it confirms that an idea will *not* work. The vast archive of zero-cited papers can thus be a huge resource for science as a whole, perhaps more useful than the few hypercited references that rapidly become part of the well-known bedrock of a field.

WHAT MAKES SUCCESSFUL PAPERS?

The higher the impact of the paper, the more widely it is intended to be read, but the fewer abstruse details can then be included. This conflicts with the mantra of science that stresses the importance of *replicability*, that a scientist with some knowledge of the field would be able to get the same results. Omitting small details often turns replication into frustration. While researchers rarely redo each other's work since it is too expensive in time and funds, and there is zero esteem acquired, they often start new research from them. Good papers give just enough detail.

How science is presented in a paper is also crucial. Great ideas buried in technical jargon and lacking motivation rarely find a selfless referee determined to bring the story to life. Scientists read just as fleetingly as the best of us in this digital epoch. Reviewers and readers alike scan first the title and initial summary, before skimming the diagrams, typically self-contained cartoons that depict the scientific story. Papers have to work on many levels to communicate well.

Science writing has to prize clarity over everything else. English recently became the universal carrier of science knowledge, through the historical fortune that science funding ramped up in Anglo countries at the same time as the globalization of science in the 1950s. However growing up using English to convey emotions, social interactions, and colloquial fireworks of culture actually makes it *more* difficult to adapt to a scientific English mode. Well-written papers praised by referees do not read like compelling novels but come clipped and linear, subclauses banished, and metaphors suppressed. Words I enjoy do not fit lowest-common-denominator Sci-glish and have to be wistfully removed. Scientists can become efficiently fluent more easily learning the language of science as a second language than learning to dump habits of a lifetime.

Communicating science is difficult and is now inevitably helped by high-quality graphics or videos (figure 4.9). Teams containing scientists skilled in this translation find a larger audience—the higher the impact of a paper the more important these graphics become. Journals looking for a wide audience find it easier if images supplied are beautiful and compelling, aspects not often felt to be the purview

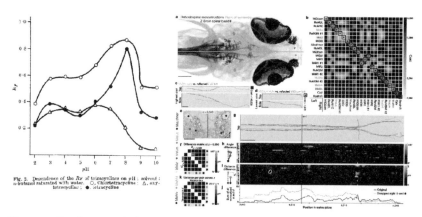

Figure 4.9: Developments in data presentation, from a randomly selected *Nature* paper in 1960 (*left*) and in 2017 (*right*). Reprinted by permission from Macmillan Publishers Ltd: Nature doi:10.1038/1871111a0 © 1960; Nature doi:10.1038/nature22356 © 2017.

of scientific rationality. Subjects such as astronomy supply these in quantity, while others such as genetics operate in a more difficult image universe. Funding-wealthy, large-scale science teams also benefit here. While compelling images increase the chances of papers being published in high-impact journals, it is still unclear whether these adornments help or hinder the growing network of science knowledge.

COMPETITION DRIVES KNOWLEDGE CAPTURE

This chapter has outlined the challenge for scientists deeply embedded in a "publish or perish" culture. Competitions is intense between journals, between editors, between constructors and simplifiers for impact (and between large teams and small teams), between disciplines, between scientific truths, and finally and most brutally between all scientists themselves. The near-complete globalization of science allows internationalized metrics such as h-factors and journal rankings to dominate previously more subtle evaluation of outcomes. The very use of these backward-looking chains of citations stresses above all a narrative of the continuous progress of science.

To trace out this ecosystem of science more fully, I will traverse some of the food chains it supports. I divide this world into five overlapping "scio-spheres" (akin to biospheres) that converge around the

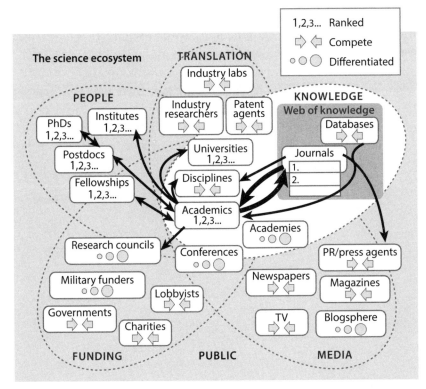

Figure 4.10: Dynamic interactions of actors (researchers and journals) in the *knowledge* "scio-sphere" of the science ecosystem. Each component experiences competition (distinguished in legend), with principal influences shown as arrows.

main activities and actors (figure 4.10): *people, knowledge, translation, funding,* and *media.* A graphical representation of this chapter shows the capture of knowledge into published archives as depicted by the arrows that link different actors, dominated by academics and their research teams, and the journals. This is generally a separated world from the outside, driven by its own internal competition. Ranking is rife in every part of it.

Emerging pecking orders are crucial to allow scientists to decide what to read, helping them navigate the information sea. As this sea becomes increasingly an ocean, the struggle to maintain connections across disciplines is becoming worryingly harder because of the rising number of scientists. The usefulness of such hierarchies is also

increasingly intensifying competition, between journals for attracting the most exciting research, and between scientists for reaching
the widest audience. I showed that this is an inevitable aspect of the
science ecosystem, relentlessly increasing the competitive pressure
whatever those caught in its snare might want. We will look later at
how this influences science success, money injected, and what gets
done. But the next part of the story is to see how scientists influence
each other, and for that we move beyond the written word, to the
gilded voice.

5

WHAT SCIENCE DO SCIENTISTS HEAR ABOUT

Scientists do not get information and insights primarily through public channels. Instead, it is direct engagement that influences their agenda, career structure, visions, and efforts. While we saw in the last chapter that electronic print is crucial, the science that scientists *hear* about is in meetings and conferences around the world.

In this chapter I explore how the conference system dominates the science world, but is increasingly ineffective. I show how this science engagement is not really under control of its practitioners despite their being completely "in charge" of it. I will consider the questions of why conferences are needed, why there are different types, and why scientists go to them. Then I will switch to discuss how conferences work, and their selection pressures, which sculpt how science is organized. I will suggest there are now too many conferences, which have become too alike, and ask how they can be challenged and be more creative.

WHY DO SCIENTISTS GO TO CONFERENCES?

THE CONFERENCE "MERRY-GO-ROUND"

Before the 1980s, scientists traveled to one or two international conferences every year to meet people in their subfield and discuss new results or directions. Nowadays successful scientists who I know hurtle off more than twenty times a year. What has driven this enormous increase in interactions, and how does it change the nature of science?

Since the rate of large-scale science breakthroughs does not appear to have increased by the same amount, this frenzy is not driven by the drama of new scientific results to keep track of. There is more science being done by more people, but dramatic results that change the view for large parts of any field do not seem to emerge more often. At most conferences few results are completely new but float a tide of small improvements to take in. Despite this, scientists are spending more time together in different continents around the world, and here I trace why that might be.

A significant driver is simply the availability of funds for traveling. Even before the Second World War, tenuous networks of scientists in different countries were encouraged to engage with each other. Institutions and individuals vied to host prestigious meetings, often using their own funds. With the opacity of the Cold War came a need to retain channels of dialogue between countries, and high-level scientists pushed strongly for an essential openness in science above all. National governments began to support the funding of scientific meetings to which almost all relevant participants, even from "enemy" countries, could be invited.

Based on this, the rise of science funded by the European Union has continually concentrated on ensuring EU scientists talk to each other. A series of "Framework" and "Horizon" programs has funded networks spread across the EU map. Stringent rules for using these funds mean a significant fraction of all the "science" money goes to holding collaborative meetings, conferences, and training schools for younger scientists. This policy *has* decreased the number of groups in different EU countries working insularly to repeat each other's results. Collaborations force scientists to think further ahead, to become more competitive internationally, and to support each other more.

On the other hand, these science funds are not for actually doing scientific research but are subsidies for airlines, hotels, and restaurants.

Increased funding for conferences means that young researchers often now attend many over their PhD training, particularly if involved in an international project, which was much rarer two decades ago. In an EU project, they will travel every six months to a new location, while in specialist training schools they devour guest lectures to accelerate their progress, fraternize with gurus, and make comradely connections with other young scientists who they will encounter again over the years. In the United States, China, or elsewhere, PhD-focused research conferences (such as the Gordon series) have similarly burgeoned. For younger scientists it is a helpful trend, but no data exists to map how this varies between disciplines or countries, or how much it helps them.

An enabler for the conference frenzy is the increasing capability to work on the move. It is often now easier for successful scientists to think deeply about their science, write journal papers, and engage with new ideas when en route, than back in their home locations with its active distractions. Although you can't wander into a lab, help a student with a spontaneous discussion, or prepare new samples, much else is just fine. Away from the chores of office and circles of friends, there is full immersion in science. This trend is particularly noticeable for theorists, who have become a class of itinerant roaming rōnin, isolated within the anonymous mass of cotravelers, gaining much work time on the move. Computing resources are remotely accessible and engagement with key collaborators just as easy. E-mailing my theorist colleagues, I have no idea where they are on their cyclical circumnavigations of the globe.

THE LURE OF CONFERENCES

The ecosystem encourages scientists by making travel funds available. Although they gain protected space to think, there are cheaper ways to do this—like working in a tranquil library or a buzzing coffee shop. What attracts scientists to so many conferences each year?

Firstly, listening to scientists talking about their work is actually a great spur. Talks are compressed in length, varying from a parade of ten minute presentations from groups working in similar areas, to hour-long keynote talks of major interest that are culminations of many years effort. This compression makes it possible to quickly extract the crucial questions of interest, or note a technique that bears further mulling and development. I can be reassured that I'm not duplicating others' work or harness serendipity by sidling into talks on subjects more distant from my own expertise, to glean new ideas and be stimulated by interesting science. I can also form a view about the strength and depth of emerging fields.

HUMAN FACES OF SCIENCE

Secondly, as a young scientist, it is revelatory to see digital artifacts (typically research papers read onscreen) transformed into a mass of argumentative humanity. It turns a local activity into a global clan—somewhere you can belong. Conferences are the first time I realize someone is actually interested in what I am doing, and who respects my efforts. They stage the journey from wrestling messy results of active research into clear and plain-speaking summaries that freeze into concrete abstractions of our work. A talk is an in-between point, where both results and the messiness are allowed to coexist. In general, the younger the scientist the more they emphasize process, and while established scientists emphasize concrete results. It is startling to hear my own work presented so differently by someone else—emphatic (if they are older) or wavering (if they are younger). So conferences are also the human face of science.

For gaining insight into different fields, they also provide incredibly rapid education. To retain a leading edge in one's research, they are a chastening reminder of the competition. Despite the vast number of potential encounters for a scientist at each, meetings only need to stimulate one good idea, or to connect with one enabling new contact, to make it all worthwhile. Each of these outcomes needs months of further work to come to fruition, while the right one leads to years of immersed research. But to collide most fruitfully, to encounter many and yet select a few, is not now well delivered by conferences.

WHERE IS DEBATE?

For all that scientists emphasize conferences as places of argument and discussion, there is less evidence it now works like this. As a young scientist I recall listening in a tribe of several hundred to a talk when, halfway through, another scientist stood up and started haranguing the speaker, who gave back as good as they were getting. After minutes of continuing bystander awkwardness, I came away with a much better idea of the problems of the theories and with much more interesting ideas, plus some sense of the vehemence of implacably held positions. This happens openly in almost no meeting I now go to, to the great detriment of science. Conflict hones honest understanding.

The compression of talks, aspirations for importance, and the decreased ethic of active participation, mean that many talks stimulate no questions. This cannot be a reflection on the science—all research generates many more questions than it answers. While one can complain about presentation of the science (both the sheer number of talks, and how well each gives an audience what they need in the time available), the general competency of scientists in presenting their work is quite high. So what has led to this reduced combativeness, or enthusiasm for public debate?

Part of the reason is that the sheer pace of meetings does not admit whimsical disturbances to their schedules. Compressing the length of talks squeezes the space for discussions. Capturing this space by asking your own question is thus a powerful act to which the powerful act. So although there is a suggestion that scientists now don't like questioning each other's authority, the challenge is really to make all present feel able to contribute to questioning. Young scientists worry about asking a "stupid" question, even though the likelihood is that if they are confused, so are more than half the audience. This barrier to entering the compressed question space dissolves when a first question triggers a cascade of others. But just when things might get interesting, it's time to move on. Speakers can (and do) preempt any inquisition by exceeding their allotted time and questions are abruptly discouraged. The most hardened chairperson faced with an unstoppably voluble theorist for instance, is helpless trying to enforce the space for the dialogue that conferences are supposed to have.

Various counterproposals for organizing meetings have been tried. One of the best I know was initiated by Michael Faraday a century ago and now run by the UK Royal Society of Chemistry. Speakers have to provide written copies of their research a month before the meeting, which is circulated to everyone. Then they are given only five minutes to speak about their work, which everyone is expected to have read, and after this starts a half hour debate. The discussion is recorded and after editing is published along with the original research. The slowly evolving debates unpick all parts of work that are less trustworthy, or open up entirely new areas of discussions beyond those initially triggered by the speaker. There is no hiding, no control of the agenda. This style of format insists on greater work for the audience (they have to read a few dozen papers before they arrive) and is incredibly rewarding for participants. But it cannot work for more than a few hundred people, or to showcase a large amount of diverse science, since only a maximum eight slots a day can be realized. Other meetings offer less extreme attempts to encourage increased involvement but without such success. As a result, most scientists confirm they really enjoy only small meetings, despite which they attend countless enormous meetings.

PRESENTING MASTERY

So scientists do not now go to conferences to listen to people *debate* advances. Most effort in their own preparations is about showing how well they have *mastered* their research, how completely they have tied it down. Rarely do they expose their current problems, their worries about the science outcomes, or the details. Mostly they try to show how they have solved all these, so a listener is led to accept, rather than be involved creatively in contributing further science. It is not surprising that participants do not ask many questions: the talks are slanted to remove both doubt and cooperative contributions from fresh minds.

One of the key reasons why scientists focus so hard on going to conferences is thus to impress their peers. They aim to showcase success, stress importance, and demonstrate leadership. Science is helped by competitive elements, but such values have come to dominate the culture of conference presentation to the suppression of many others. Showcasing of inconsistencies, confusions, similarities, and nebulous results is rare. Young scientists are impressed and appalled: it all seems so persuasively successful, so impossible to emulate.

From the biological perspective, this activity resembles the fantastical displays made to woo potential sexual mates (of either gender). Scientists with more dramatic displays rise up the pecking order, showing they are better adapted to dominate the scientific landscape with their memes. This is also directed at attracting funders and indirectly society at large. As science generation begets generation we keep the fitness of the whole improving this way. Such excessive displays can aid our selection within an abundance of choices. Just as in nature, this is not efficient in resources but it inescapably arises from competition. Each iridescent butterfly of an idea strives for the most dramatic wings to flash sunlight-flecked colors into the furthest distance, hoping for a better mate. One can imagine that in this landscape, little effort is placed on any sterile mating between different disciplines that do not have a connected pecking order. Hence the tendency for continued separation between physics, chemistry, and so on, as well as their subfields.

In this vein, scientists will not turn down the opportunity to tell someone about their research, as it is increases their visibility and worth. Since few scientists have time to read all the papers coming out in their field, they judge the value of a colleague's work from their talks at conferences, and on how many they have given. Conferences have become essential in the valuing of scientists, in sorting the pecking order of their tribes, and so for careers and opportunities. Scientists are thus more in transmit than receive mode and have adopted practices such as flying in for a day to give their talk and then leaving for the next meeting elsewhere. Besides slicing away their opportunity for interactions, particularly with those they have never met before, this decreases still further the potential for debate. This is a real loss since research leaders are actually needed not just for their own research, but for their views on other scientists' advances.

LIFE ON THE CIRCUIT

Perhaps one driver for conferences is their transformation from monastic sacrifice to business lifestyle. Conference schedules expect scientists to travel over weekends, accessing cheaper airline tickets (available because businesses don't insist employees do this). Time for relationships and social activities is thus deemed expendable. But the same scientists impose this, both women and men, since they do the conference scheduling! When I challenge any particular conference committee, there is a feeling of powerlessness for changing the culture. There is no union for scientists to challenge such norms, and little open discussion. It says most about scientists own view of the stereotypes of being a scientist. Perhaps they have bought into the myths of individualistic, single-minded, and relationship-suppressing people, despite their own experiences. Interestingly, this inflexibility of the science ecosystem seems to be gender neutral, since research fields that are either female- or male-led run conferences similarly. Reflecting all this is the lack of evaluation, asking what the payback has been of attending each conference.

WHO DECIDES ON THE CONFERENCES?

We have discussed scientists as passive attendees at conferences. But they also act as the instigators, both through small groups persisting in self-organizing committees over time, and through learned societies that aggregate very large numbers of scientists' interests. Why does anyone decide to put on a conference?

SCIENCE COMMUNITIES

Conferences tend to run in continuing series, spaced every year or two apart, which allows them an essential role to build science communities. Subfields in which a set of researchers meet regularly every few years tend to thrive on a whole variety of levels. By keeping track of related scientists' progress, research is optimally directed to new ideas. By forming taut thin networks of senior scientists, high levels of trust are created, which can open up more dialogues and greatly enhance cooperation to improve progress. On the down-side, such clubs can exclude those challenging them. However, younger scientists are visible early in their careers and are mentored and fostered by the older researchers they repeatedly meet. As in human societies, when interactions become impersonal and the communities become too large, reciprocal support disappears. The hollowing out of our social communities is now replicated within science communities.

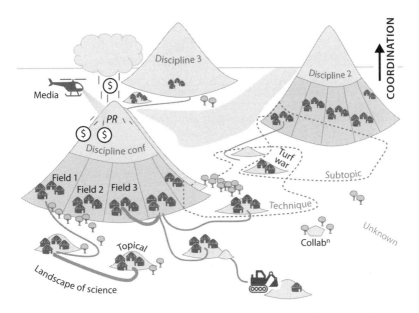

Figure 5.1: The landscape of scientific conferences contains natural and man-made barriers, and groupings from major disciplinary jamborees needing high levels of coordination, to small topical meetings.

Digital social networks do not yet provide the answer because we find it hard to relate closely to ever more people, or mentor people who we do not know.

In young fields, those with exploding numbers of researchers, or in widely interdisciplinary fields, building communities is extremely difficult. No single conference can cover the whole of a very large or very wide field. Instead many small, fragmented meetings are put on with focuses that partly overlap and serve different communities. Only if the overlap of people between these different meetings is large enough do ideas and techniques percolate across the valleys and rivers (a landscape is indeed good way to think about this; see figure 5.1). One example is the area of nanoscience, which studies new properties of matter emerging at tiny size scales, and which is so broad as to include every discipline but ends up being fragmented with little sense of community. This major problem holds back its development, since ideas seep across only slowly despite the novel and intriguing results.

Typically each science community develops a series of conferences, to help foster and enhance connections. The expansion with time in the number of conferences reflects several trends in the underlying science communities. First is the rise in interdisciplinary research that requires new communities to be built and serviced by conferences. Second is the increasing number of scientists overall, coupled with the diminishing returns available from the larger conferences, which leads to proliferation of smaller meetings. Thirdly, actions by nations and regional pacts (such as the EU) to encourage communities in strategic areas lead to yet more meetings. Perhaps the best news behind the increase in conferences is the evidence it provides of tangible communities.

BIG IS BEAUTIFUL

The dinosaurs of these communities are the scientific societies, large organizations many centuries old, with professional staff representing tens to hundreds of thousands of members. Allied to traditional disciplines such as chemistry, materials, and microbiology, these societies hold conferences striving to unify their fields and attract tens of thousands of attendees spending five days together. Such meetings are so embedded in the culture of these societies that their structure and scale has been frozen into perpetuity. Their unwieldy size is growing even faster than the number of scientists, at 6 percent a year for some, which is again not sustainable.

Clearly not everyone wants to listen to the same talks in such vast extravaganzas, and so they are split into very many sessions (sometimes exceeding a hundred at once), each operating to some extent as their own miniconferences held simultaneously. One of the most common sights in such a conclave of researchers is to see scientists running through vast carpeted corridors from one room venue to another, conference bag slung around shoulder and hair trailing behind. Deciding which talks to go to is a major challenge as they are often grouped in peculiar ways that require energetic translocation. It is frustrating that the talk in the neighboring room always seems more interesting than the one I happen to be listening to.

Perhaps the loneliest sight is glimpsing an enthusiastic researcher talking to a scattering of only three people, sprinkled across forlorn ranks of padded seats, while thousands scurry past on the way to some other attraction. Only sporadically in the program do major sessions bring larger numbers together for prestigious talks. Seeing the weight of interest in the different sessions gives a democratic snapshot of what is hot, trendy, or just breaking.

Sometimes I privately duel in a hallway with a researcher who doesn't pay attention to my published works, or a researcher who is similarly sounding me out, forcing their views on me. Sometimes it is a scientist I would like to collaborate with but am not sure of their trustworthiness. Finding someone who shares your view of a corner of research, or who can enthuse with you about promising directions, is reliably rejuvenating. I believe this is why scientists really go to conferences, for these few moments of clarity through interaction. In large meetings, such moments are few.

Another downside to conferences is their inevitable lubrication by food and drink. Short conversations require coffee, longer ones lunch, neither ideally suited to invigoration (compared say to the remote moorland walks of my childhood). As a society, and as conference organizers, we pay too little attention to ways of being together that don't involve consumption. And to creating the spaces that science really thrives in. These large discipline conferences thus bring entire fields together but have become so big that they fragment, and mainly fail to nurture spaces that create communities.

THE OTHER END

At the other end of the conference scale are small meetings of collaborators or network meetings, up to a few dozen people. These can be a joy or burden depending on their sociability. But the funders' drive to enhance interactions has led to many more of these meetings, often convening in new locations every six months.

To evaluate the success of such a meeting, it is only necessary to listen to dinner conversation. If it concerns using electric eels to electroplate gold, or inflatable telescopes in the upper atmosphere, or any other element combining fun with science thinking, then this is a meeting that will deliver creativity. Other meetings may be essential for reaching goals in a large funded research program, for aligning different components of research, or for weighing competing approaches, but not be inspiring.

Why are so many science meetings dutiful, not creative? And why do people organize them? Wrestling the bureaucracies who demand frequent meetings is fruitless—for funders they are a way to measure success of an initiative. Too little time is spent reflecting on organizing inspiring meetings, to create the sense of open space that is the most useful. Scientists are not very clear on their aims in a small meeting, as "open space" is not a very tangible output, but a by-product. Training the organizers to challenge participants would be a step forward.

CO-OPETITION: COOPERATION VERSUS COLLABORATION

At such small meetings, the best opportunity is to form strong relationships with other researchers. A colleague-of-a-colleague style of aggregation knits new connections, with great opportunities for embedding trust and scientific empathy. Trust is a huge issue for scientists, since often their best work is founded in discussion with others, and takes much time to gestate and emerge in appropriate form. I need my colleagues to keep it under wraps, but without their necessarily being directly connected or benefiting from it.

As public funders ourselves of these conferences, society might feel that concealing ideas is against its interest. Perhaps all ideas ought to be written up electronically at the earliest stage, free for anyone in a good position to make them happen, find their truth, or trash their naïveté. However while ideas flow easily in science, realizing them is extremely difficult—a list of interesting ideas is not as useful to the science ecosystem as a set of grounded truths derived from working through them. Scientists gain cult status by rendering new truths to a smitten audience. Allowing them to hoard their nascent ideas, discuss them in small friendly cabals, and drive the most juicy ones forward seems to get most back for our science investment.

Discussions about this balance, or "co-opetition," have emerged in discussions of social entrepreneurship. If a business is looking to improve society in some way, then giving their ideas to competitors might actually be good, even if it drives them out of business. But this view sees developments as decoupled from the ego of humans, and competition as not always being helpful. The challenge of growing science is to avoid privileging competition over sharing, but both are needed.

Scientists are highly aware of this balance between open and closed circulation of ideas. Their need is to be personally identified with good ideas and the competitiveness of current science means that it is risky to wait too long before talking about a not-yet-realized idea. Conferences are ideal forums for this, since the presentations are informal and unattributable. Frequently the best ideas emerge spontaneously from a discussion, not from sitting in an office thinking alone. Being forced to articulate on the fly, defending an idea, and sparring with extremely smart people stimulates the mind to fantastic flights and concepts. So we improve our ideas by talking to others, but also release them to fly freely.

Small meetings thus provide better opportunities to bond and brainstorm than the discipline jamborees but vary enormously in their liveliness and in their creativity. Frequently they are driven by funders' demands for compliance, rather than science directions.

THE IN-BETWEEN OF TOPICAL CONFERENCES

Between massive and tiny conferences, it is more normal to attend something in between (figure 5.2). The distribution of participants in meetings peaks broadly around a few hundred scientists—often a "topical" meeting of scientists competing in similar areas, bringing

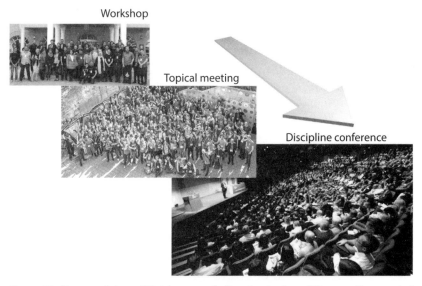

Figure 5.2: From workshops ("Driving towards Decarbonisation of Transport"), to topical meetings (7th International Conference on the Science and Application of Nanotubes), to disciplinary conferences (American Society of Clinical Oncology Annual Meeting). Photos: Habib M'henni/Wikimedia Commons; Jason Krüger/Wikimedia Deutschland.

representatives of many groups from around the world together for a few days argument. It is pragmatically manageable, with only one or two parallel sessions of talks, everyone eating together, and a sense of closer community. It is no accident this is also typically the size of a tribe.

Robin Dunbar, an anthropologist from the UK, suggested in 1992 that extrapolation from bands of primates implied the maximum size of any society in which people knew each other and their place was 150. Beyond this, laws, restrictions, and norms are needed to keep a cohesive group together. Groupings dependent on collaboration for survival, such as Neolithic villages, Roman army units, or successful companies, strongly favor the 150-person scale. Even your number of active Facebook connections tends to this number. This seems to be also the typical size of an effective academic subdiscipline, and close to the maximum size of a useful conference, though neither have been well studied. It is not so clear what happens if subfields don't fit this size.

In a topical meeting, the sense of being in a community is enhanced by the expectation that everyone shares a common background, so that introductory material doesn't need to be reprised. Speaking about a year's work in half an hour means cutting to the chase rapidly and the density of information is high. It is a rewarding way for scientists to learn, often much faster than reading the research papers on which the work is based.

WHAT FITS INTO A TOPICAL MEETING?

Another norm that has evolved is the maximum length of any meeting. You will get scientists for a week, but any longer and they are elusive since demands of teaching or organization conflict. The topical conference strait-jacket is thus set: in four and a half days, two dozen specially invited speakers are given half-hour talks. This leaves maybe fifty others each speaking in tight fifteen-minute slots to fit into the program. Competition for airtime is thus intense. An audience of 150 can fit in a room and feel communal enough to engage in some sensible discussion. Any larger and no one asks questions, the buzz drops, and science takes place outside the conference hall instead of inside.

Scientists thrive on being heard, not just hearing others. To persuade scientists to attend it is essential to give them a platform. With seventy-five speaking, there are another seventy-five shifting about impatiently in their seats. To finesse this, the other traditional conference forum is the poster session. Turning last year's work into large cartoons of pictures, diagrams, and text, contributors create a glossy board-sized poster. A room full of these display boards becomes a hubbub of noise as those not tethered to their own posters browse around, choosing particular researchers to argue with. It is a great opportunity to really challenge someone (who can't go hide) on how they got those results, exactly how they set up their experiment, or why you don't agree with their conclusions. Onlookers to these debates act as both observers and appeasers, but of course only fragments are remembered.

A snapshot here is the silent imploring young scientist marooned at their poster, while passers-by flick their gaze rapidly over dense, small-font text arrayed from top to bottom and then hop rapidly past to something more obviously engaging. It is like being a shopkeeper imploring customers to buy at least something, but watching them all drift by. No reconfiguration of your printed poster is possible on the fly, though large flat-panel screens are upping the ante and allowing videos and all the elements of active advertising to emerge. I expect slick imagery, virtual reality, and targeted science commercials to follow.

HOW DO TOPICAL MEETINGS EVOLVE?

What happens to such meetings as more science is done? If they stay the same size they become increasingly competitive to attend, or otherwise they fragment. Subfields, the owners of topical meetings, try holding onto previous research areas even as new ones emerge. The intense competition for airtime to present at conferences gives strong incentives for budding off the newer sub-subfields, particularly if increasing numbers of researchers become interested. Scientists from this new area with pretensions to act as leaders will jump on any chance to start a new topical meeting, and that in turn is possibly only because of the many ways they are financially supported.

Most of the behemoth learned societies provide simple routes to set up a topical meeting, supporting some of the infrastructure and advertising to the community. As long as you have a good science program, then it is likely to be a success for these new organizers, pulling in enough scientists to break even financially, and raise their international profile. Sometimes national or local considerations are important, since officials pride themselves on being able to showcase their locality to esteemed international scientists. The scientist acting as local organizer will garner kudos for their international reach and world standing. It is for this reason that bemused scientists attending a conference opening session are often subjected to politicians lecturing them on the local economy or high-technology leadership. Every country seems to have aspirations to be a "knowledge-led economy" and espouses this mantra while doing basically the same as all other countries. Needless to say demanding questions are not part of this conference opening session.

Once a topical conference has been running a few years, it tends to acquire a tenacious life. Birth is always tentative, with a few participating in its handmade construction. Before they know it, an emerging community has expectations and feels committed to attending subsequent meetings. The ring of leaders expands, as authority and responsibility expand outward to share the burden and culpability. A scientific advisory panel will become associated with the topical series, combining colleagues of colleagues with representatives of fields. Their principle seems to be "onward to the next conference," and financial survival. Woe betide any scientist (like me) who suggests that perhaps there are too many conferences in this area, and this one should die. Conference suicide is outside any morally tenable discussion. The only routes are merger, subsumed into a greater mass, or fission into divided meetings of youthful dynamism.

None of this is well thought out, but instead a spirited aggregation of aspiration and pragmatic hard work. Merging topical conferences is inevitably unhappy as the 150 barrier is breached and an unmanageable bloated hybrid is created. When topical conferences fragment, all participants have to choose which part to abandon or whether to expand their meeting schedule. So it is the inexorably increasing number of scientists around the world that leads to these choices and results in the vast increase in conferences. The lack of their natural deaths prefigures how further increases in human lifespan will affect us, when more is not necessarily better.

COMPETITION OF CONFERENCES

Most unhelpfully, the pool of conferences is not subject to much competitive pressure. Scientists gain by being organizers of meetings, and promotion or prizes may depend on being seen to be heavily involved in such community leadership. As a result, it becomes important to make such conferences successful and make sure good science is represented. Twisting arms of highly reputed colleagues into attending is one of the real jobs of organizers. Killing a conference would remove esteem from a field and from the current generation of its motivators.

While there is clear sense of prestige attached to publishing science (as we saw in the last chapter), that attached to giving conference talks is fuzzy. Gaining the opportunity to talk at a massive-scale conference has high worth, but scientists know opportunities to discuss new science here are lower. Learned societies are happy when their conferences increase in size, reporting this as their enhanced weight in science, and bestowing them greater presence in national lobbying. Topical meetings are instead where the peers you really want to impress are, but they proliferate continually.

Killing a conference also "disinvests" science funds from businesses of a host city. All meetings are now encouraged to evaluate their local "benefits." Typically they use an "event impact calculator," in which the money brought into a city by the descent of scientists into airports, taxis, hotels, venues, restaurants, and bars is added up. Such calculators (encouraged by city halls and trade organizations) make no attempt to balance these against the other side of the equation: lost time, poorer management, environmental impacts, or the true benefits to society. More to the point, there is no sense of truly evaluating the impact and benefit of conferences—it is yet to be built into the science ecosystem at all.

Underpinning all this is the support of funders to underwrite the activity. Overall increases in science funding are not connected to any careful scrutiny of the percentage spent on conferences. Funders assume that being invited to talk at conferences is a good outcome, boosting the need for researchers to focus more on them. Nowhere in the science ecosystem is there any explicit attention to how many conferences it is worthwhile to attend. While it is possible to argue that even one rewarding encounter is beneficial, this has to be set against the opportunity costs. Funds can be spent elsewhere, and most importantly, time can be spent on other activities.

This lack of self-scrutiny means that no one even collects statistics on who goes to how many conferences (though see the website). Current measures of a successful conference are: did it break even, did enough people come, and will they come again? Circumstantial data on the best size for banquets, the length of poster sessions, and how much food or drink they should be accompanied by, as well as much else, is left as implicit convention. Meetings without long-term leadership bounce from style to style, while the yearly megameetings find it hard to avoid ossification. Myriad opinions rather than any consensus suggest the "right way." And none explicitly focus on engendering those magic moments that set researchers alive.

Let us imagine that researchers instead spend one of their conference weeks each year differently—collectively discussing recent research directions. Assigned collaborations, committing a week of time but virtual rather than requiring travel, could produce a joint review reflecting on advances in a subfield. This would be accessible by others, use electronic media and informality, exploit group views rather than individual polemic, and help guide researchers through the web of emerging new published material. Key is to find a way to reward individuals who develop such activities.

In the end scientists sustain the current system because alternatives cannot yet be envisioned and funded. Conferences can still deliver the "zing" of new insight, creativity, and problem solving, but only sporadically. They also form the lifeline of connections that keep the whole ecosystem bound together, albeit with increasingly shorter lengths of tighter-lashed ropes. As a result the only choices scientists consciously ponder as consumers is the selection of which meetings to go to.

WHICH CONFERENCES DO SCIENTISTS WANT TO GO TO?

Homes give us safety, secure our assets, and are where we let our hair down. But how many conference homes can a scientist have? This is an ever more difficult question, as the number of subfields and conferences that might sustain each scientist grows. At the start of their careers they get little guidance, and the practice within any subfield varies widely.

Young researchers drool over announcements for conferences in exotic locations, talking up the esteem of this meeting to their boss. One of the joys and perks of being a scientist has always been the international interconnections. But since the cost of travel, hotels, and conference fees is typically several thousand dollars, any research head will be cautious about committing funds unless the reason to attend is compelling.

For young researchers who just completed their first project, this is their opportunity to discuss it with other scientists, make connections, and feel for the first time part of a field. Rather than a panic-stricken outsider wondering why everyone else is so much smarter, they can attain the feeling of belonging and acquire a reason for all those dazed late nights and enthusiastic weekends in the lab. Research supervisors like students to go to conferences as soon as they have results, to motivate them (to work harder). Partly this can show students just how competitive science is, and how people all over the world may be working on the same problems. Students hopefully return with a much wider view of research and start to think in their own way what good directions to pursue.

It is at the next stage of career, where a researcher is in charge of their own funds, that things become more complex. Repeated contact with colleagues and competitors in a topical subfield forms a stable tribe for them, and they can see where they stand within it. Once they find a group of scientists friendly and inclusive, they will be happier to go to their meeting again, as long as the others also keep attending. The most important asset for meetings then is their aura, the atmosphere they manage to engender, which becomes associated with the meeting brand. Like a microclimate formed by a biocommunity, its assets are its sanctuary and stability.

As a scientist's work becomes significant to others, they may find themselves "invited" to present at a meeting, rather than submitting the short summary abstracts that are ranked to bid for time slots. Invited talks are markers on a CV, indicators of research esteem. Compression of scientific presentations into the scarce available time means high competition for such longer talks. It is hard to turn down the offer of giving an invited talk, and it becomes still harder for the next levels of esteem, "plenary" and "keynote" talks, normally given to the massed attendees. Researchers thus compete to access the most prestigious conferences. But as we shall see, conferences compete to access the best researchers.

LOCATION, LOCATION, LOCATION

One way scientists choose between conferences is where they are held. European nobility have donated many astonishing palaces and castles to universities and foundations for use as venues, motivated by their realization that investing in the future is the only powerful legacy. Scientists always find some rationale to organize new meetings in such stunning locations, relentlessly adding to the number of conferences.

Carefully constructed venues combine the lure of novelty with subtle transportation problems that prevent attendees from disappearing to the brilliant beach instead of darkened chambers. Spectacular Italian hilltop castles miles from anywhere, or remote mountain lodges, are spectacularly effective. All scientists accumulate stories of travel quests progressing from airplane to goat cart, of baggage that arrives just as they leave, or being stranded in unexpected places. During the 2010 Icelandic volcano lockdown of the upper atmosphere, my students called from Mexico to ask what to do since they couldn't get home from their conference. I had to convince them to find something to explore rather than stew in their rooms waiting to get back to their labs.

HOW TO CHOOSE A CONFERENCE

Scientists also choose between conferences on the basis of interesting invited speakers. If I know all their names it is not such a good sign—perhaps I only hear the same people time and time again, so learn little beyond reinforcing group-think. If I know none of the speakers, then perhaps they really aren't doing anything terribly exciting. It is very hard to guess if I will learn something important from a conference before I go, or experience stimulation and creativity from an open exchange of ideas. There is no openness index for meetings that I can check.

Sometimes, the pressure to attend comes from the dangers of being missed. Will I no longer be seen as a significant actor in the discipline? Will I not be party to conversations that build a mutual support club? Will I not be "important"? Competition in science intensifies this endless participation, ratcheting up as competitors speak ever more.

Even if I am invited to a conference, rarely will my costs be fully covered by it. Proselytizing scientists will mostly be supported by their national governments and taxpayers. The better they are, the more they'll travel to conferences, but the more difficult it becomes to find enough time for them to carry on their best research. Thus they pare down time at each conference, depriving younger attendees of debates and serendipitous encounters. Invitations come weeks or years in advance, and it's hard for scientists to decide which to accept, harder still to decide for those who want to privilege their own personal relationships. Such selection pressures unbalance the ecosystem by elevating only certain personality types, working against its overall interests.

With more conferences, the need to spread a message without it drowning in the sea becomes more important. These are the factors that propel scientists around the globe, and despite widespread disquiet about carbon footprints and unsustainability, no alternative model has traction.

HOW DO SPEAKERS GET CHOSEN?

Having looked at what drives scientists to conferences, we now look at the conference ecosystem. Each conference aims to bring prestige to the organizers. Location, celebrity speakers, established clientele, and a good brand are needed, as with any events. Organizers have to build an attractive program early enough to convince their potential audience to earmark the meeting dates and identify funds to come.

A scientific advisory committee defines the scope of the conference and the range of subfields represented, setting how many people would be interested. They then suggest names of potential speakers for the invited, plenary, and keynote talks. A single keynote talk for the conference, for instance from a Nobel prize winner, and a plenary talk every day from a highly successful established scientist, work well to keep delegates attending for the longest time, and rouse them early each morning.

Invited talks delineate the range of subfields, ensuring groups send other delegates too, since invited speakers have a vested interest in making the conference a success. Targeting of larger audiences leads to inventing trendy sessions to capture the imagination of potential attendees. Since new areas do not often have many results yet, such sessions can rapidly become clones of each other.

CARTEL CAROUSEL

How does this committee of scientists choose speakers? Often they take inspiring talks that they recently heard at another conference. This is very prevalent but leads rapidly to overusing a cartel of famous speakers. The selected few hop from conference to conference, while their research itself progresses only at the normal rate. One has heard their talk before, and while a good show, it ceases to stimulate new creativity. It also has the effect of driving other researchers mad to break into the cartel.

Once in the cartel, if a scientist brings in the crowds and has the golden touch then, as in mass media, they become the figure of the moment. Their name on the conference advertising gilds any meeting, and the invitations build up thick and fast. At this point many a speaker regrets the position that they gained, feeling closer to entertainer than scientist. Good speakers start to wonder if they can say just anything and still be adulated. Questioning is mostly timid in the face of such reputed greatness. Of course, if the science really is dull or weak, the crowds will drift away, and in time subfields trail off into the devoted few acolytes. Harnessing the cartel carousel primes the success of a meeting. It can then be used to reassure potential meeting sponsors of a high profile, and allow a secure financial budget to be put forward.

Another tactic for selection is by word of mouth, discussing with colleagues who they would like to hear. Program committees are chosen to give some balance across different countries, genders, or experience. Members are often partisan, focusing on their nation's figures who have been doing strong science. But in many fields it is no longer easy to know who has been getting interesting data. Researchers even in a single country are disconnected from each other since they rarely attend national meetings whose scope is necessarily

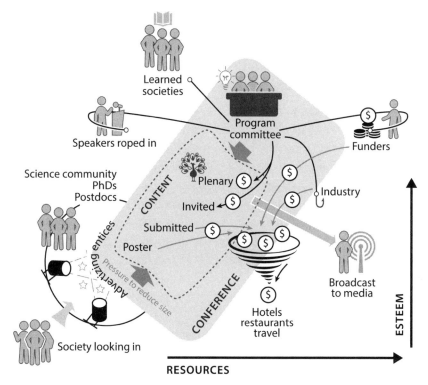

Figure 5.3: The conference model and its pressures.

smaller than international meetings. A frequent but bizarre experience is finding time to talk to a colleague working in the building next door to mine, at the conference on the other side of the world we are both attending.

As with innovation or in any ecosystem, it is hard to introduce novelty because customers and backers are risk averse. Most conferences are remarkably similar to each other (figure 5.3), not because the model is perfect, but because the system crushes out change. A disappointing aspect of topical conferences is the way they can develop into a closed group of eminent scientists who persist on inviting each other. There is a tradition in many fields of preventing committee members from speaking, however their protégés are often substituted. Such insular communities persist for a long time, their stability producing a comfortable support system, but reducing the intensity of competition and sometimes the pace of the field.

INDUSTRIAL CREDENCE

Many fields of science are some distance from practical technology. That can mean they revere industrial researchers who tangentially work in the field, and entreaty them to give talks. Participants appreciate hearing aspects of their research interesting to industrial scientists, who in turn profit well from retaining interest in more fundamental aspects of their subjects, and bringing ideas back into their industries. In more application-based fields of engineering or biotechnologies, company scientists are treated cautiously, since they are in direct competition for new techniques. But any willing to invest time in conferences become part of the carousel of omnipresent speakers.

MINING DIAMONDS

Deeper than this selection by word of mouth or by celebrity is "mining," looking back through the last year's journal papers in a particular field for interesting contributions. This is really time consuming, but it greatly broadens the vigor of any scientific meeting. While tractable in some subfields, in many others once I start this task I rapidly become distracted, getting lost in a maze of new science. The flaw with mining is the shock of finding my selected speaker is not a good communicator. There is nothing more disappointing as an organizer or participant than finding the talk that I was looking forward to delivered in a dull monotone, rambling and poorly paced. There is no ranking metric for speakers.

Often I feel that I could give a better talk on the science myself. Occasionally this happens when a speaker cancels at the last minute, and persuades a colleague to give their talk. Greater distance and emotional detachment mean that stand-in speakers convey key points at the perfect level for the audience. Better conferences would flourish I suspect if everyone had to deliver their competitors' papers. They would cover both good and bad points more effectively and limit their polemic and critique to a social norm that emerged. Rather like a good book review, which does not just dwell on the specifics at issue but ranges more widely over the subject material, a less partisan view of the science would result. Many less-than-confident or short-of-time speakers have wondered if it is possible to employ a talented actor to give their talks in their stead. While this would be completely against acceptable norms currently in science, it could perfectly well result in better dialogue and stimulation. Questioning could still be with the scientist concerned. And it would differ little from having less experienced students give the talk, as happens now. The main block is the funding and training of such presenters, who would develop their own cultural framework.

NO REJECTION

Once a topical conference has selected its showcase speakers, it aspires to attract numerous researchers to come and listen to them. On the final abstract deadline day, submissions rain in. Several hundred or more are sorted by the program committee into those that will be given oral slots, and those others that will be given posters. Papers are rarely rejected outright in most fields, because this would result in a delegate not coming (losing income for the meeting).

Customers do not select based on the conference fee level, which have become fairly standard. Some conferences aim to make a profit, either for a professional company, or for a learned society (who invest in young scientists' careers). However since scientists themselves normally run the conference, there is an accepted level of fees and subsidies to help balance the community of attendees. Subsidies are frequent to scientists from countries with low levels of science investment (Africa, some of the Middle East, some of South America) and early-stage researchers. Such investments have been crucial in ensuring connections between scientists across societies in conflict. With shame I notice how rarely is a conference attendee from anywhere in Africa.

Conferences work through a peculiar market, sustained by donated resources (travel is mostly funded by taxpayers). Take away one meeting, and perhaps half the participants would not substitute it with another (all the invited speakers). Increasing the supply of conferences does not lead to enhanced competition that makes the price for each event drop. Instead it is possible to add more and more conferences, but only until a critical point at which travel funds for all possible participants are completely exhausted, or scientist's time is saturated. Because conferences have become international stages, the available pool for participants is very large, and since rewards for attendance are large (in esteem), support from developing and newly emerging countries remains strong right to the edge of this critical point. This inelastic demand (where price doesn't depend on supply) resembles a "necessity," which it is for scientists in their ecosystem.

SYSTEM VERSUS INDIVIDUAL

The conference explosion is driven by benefits to individual scientists, who invest with their funding and time, and are tempted or coerced by their colleagues. For the ecosystem as a whole, conferences form an essential way to distribute new information across fields and communities (figure 5.1). The giant discipline-based meetings cluster aggregates of topical meetings, all in a halo of related areas. Topical meetings most often are owned intellectually by a specific discipline, defended hills in the landscape. Finally we have the small meetings, often of collaborators, which can most easily connect people from widely different disciplines focusing on a particular vision or challenge.

How fast does knowledge propagate through such a network? This depends crucially on the diversity of people at each meeting, and how many different fields of topical conference they attend. Typically the span across the network is limited since scientists are mostly specialized in a subfield, and link only to closely related areas. Conferences without transient visitations from other communities can act as barriers in the network. This situation is a type of "percolation" problem, harnessed by the PageRank algorithm used by Google to identify webpage connectivity and hierarchies, or similar to how social mixing can happen in our societies.

Subtle behaviors arise because the percolation of knowledge does not always improve when adding more topical conferences. Since mixing remains local (the cartel of speakers at each meeting remains too exclusively from the limited pool of the subfield), additions can enhance local circulation of knowledge but limit global distribution. The saviors of knowledge distribution are small collaborations bringing people together from very different knowledge bases and subfields, spanning with a single jump right across the knowledge space. The diffusion of knowledge out from these sparsely populated nodes is slow but introduces critical bridges.

Building such collaborations is good for the ecosystem, but not rewarded for the individual. While payoffs can be huge, potentially inventing entirely new disciplines, the risk level is extremely high. Of all the possible thin, long links across our network, only a few are likely to be productive, give new insights, or open up new intellectual spaces. Esteem is not conferred on such partnerships, but more often puzzlement. Since such links tunnel right through the halo of knowledge surrounding each discipline, local onlookers see only part of the picture and diminish its value. Supporting such collaborations is very much in the interests of funders, as we shall see later.

How can knowledge diffuse faster? Why don't scientists ever attend a meeting completely outside their discipline? Once again the individual rewards are not high. This is not the peer group to impress whose influence will help you later. These people seem to speak a different language, and so it is rare they stimulate extensions of your current research. Often they urge dropping it entirely in favor of something new from their own area. Topical meetings will sometimes attempt a mixing, seeking speakers from wildly different domains tenuously linked. For the speaker this is both exhilarating and challenging. Everyone is interested because it is shiny new for them, but they can rely on much less background knowledge when making explanations. The role is ambassadorial, with more interest in summarizing the subfield you come from, than in the specific research you have accomplished yourself. Few scientists find it so easy to let go of their own contributions like this, but rewarding such ambassadors should be encouraged.

In our ecosystem, the role of topical subfields is for self-protection, like the safety of herds or prides—altruism and mutual support work well across science. They function to support the bulk of members, who don't all need to be the very strongest or genetically fittest. In each discipline, more scientists will survive than just those who do the very best science. Such ecosystem features may not be the optimal way of organizing. Perhaps we need predators that usefully cull, or viruses to sweep away a cohort susceptible to extinction, clearing open space for young science saplings. Instead we have large-scale funding changes (in our metaphor, shading of sunlight), that mainly entrench the status quo and, as we shall discuss next, fashions that balloon up co-opting extra resource.

THE CONFERENCE *SCIO-SPHERE*

I have introduced many of the tensions and forces that created the culture of conferences dominating science and scientists' lives. Continuing expansion of the swath of meetings is inevitable given these pressures, but counterproductive. Scientists make the meetings, but meetings make the scientists careers.

Conferences are governed by the norms of a community with dispersed control (figure 5.4). Speakers, attendees, committees, scientific

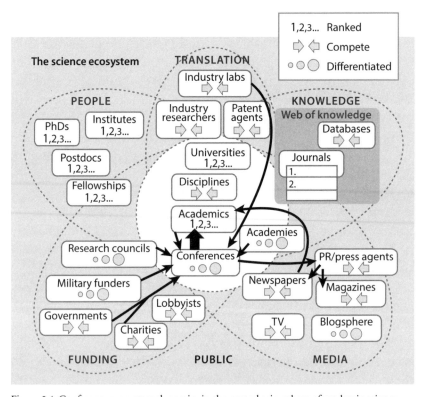

Figure 5.4: Conference ecosystem dynamics in the central scio-sphere of academic science.

societies, conference enterprises, and funders all mutually influence each other and are complicit in its ecology. However, a lack of introspection exists around this whole activity, from funding to outcomes. Changes would impact the entire ecosystem, and I noted several opportunities that exist for positive approaches. This landscape is important, because the way that knowledge moves through science is the driver for the value it creates in our societies. I have suggested that more attention be paid to fewer better-conceived meetings. In any case, this is an area where competition does not seem to effective in driving improved quality, and no metrics yet exist.

Scientists use meetings to learn what is hot. Their funders use it as a proxy measure for success. But how does the rest of society gain some insight into science that it funds? In the next chapter we will consider what tiny fraction of the science being done makes it into the public domain, and who is choosing it for us. We will find a yet more competitive landscape awaiting us.

6

WHAT SCIENCE DO YOU GET TO HEAR?

With eight million scientists working on different projects, how is the very small fraction that comes to our attention, something less than a millionth of a percent, being chosen? A significant trait of the science ecosystem is the way that science outputs are filtered and sifted, in neither a straightforward nor a random fashion, and it is to this that I now turn.

There are many layers of invisible filters at work, woven on the loom of the science ecosystem, by actors with different motives than those creating the science. This filtering is not random selection, and we should understand in whose benefit it is made. We will see that feeding our demand for entertainment risks obscuring the real goals of science, and the rationale for funding it. Our view of constructor science is becoming comprehensively skewed toward only directly applicable motivations. This is one of the reasons that scientists worry about keeping tight control over what science gets done. In looking at what rises to the surface from the depths of science, we will also discover a set of features that boost any science story, a tool kit that encompasses hooks and triggers, patter and money, images and controversy.

WHAT SCIENCE DO WE HEAR ABOUT?

We all have some interaction with science. Many people have a residual interest sparked by encounters as kids, which they now nourish by reading newspapers or magazines, watching documentaries, or stumbling on net stories. The media have picked up on this interest, which is increasing in all cultures, and have responded by investing more care and resources in reporting science. Estimates suggest that a fifth of all Americans read some science reporting, online or in print, each month. In many societies teaching of science at school has mushroomed to meet needs and desires to understand scientific ideas earlier in life. Science is reported all around us, reflecting the fruits of the significant investment by our cultures in research.

What science do you yourself want to hear about? A particular research area—do you know vaguely which research areas are currently hot topics? Or do you want to hear about the "important" science—science that might change your children's or parents' lives? It might be you want to know about what might soon change your own life. Perhaps you just want to hear science that is entertaining, and provokes you to muse in different ways, as stimulating as a museum might be. Or you want to hear about science that you paid for, because it is publicly funded in your country or through the charity you contribute to.

Each of these could motivate a different cross-section through the strata of science. Just as your newspaper or website selects news stories that accommodate to your own view of life and interests, science reporting winnows down from any number of possible stories into those few that "fit." These are the ones transformable into narratives that will get across the essentials to you in the time you are willing to devote. There can only be a few, because most of us don't put in much time at all.

Since taxpayers fund science, we might expect *all* the things that we fund should be turned into stories that we could choose to read or not. But to turn science into fascinating stories is expensive—even a few paragraphs costs $1,000 or more. Part of the problem is that scientists themselves are not the best people to concoct this transformation. Over many years they have developed ways to discuss and convey their science to each other, but these are idioms that the public hasn't similarly developed. It's like the confusion from listening to political conversations in the foreign language in which you can barely check into hotels or buy food. The way a scientist has trained their mind to flow is not conducive to creating translations into good journalistic prose with ease. What one needs are people who *have* trained their minds to make this transformation effortlessly, both skilled at reading through the science and skilled at relaying it without losing the core essence. They see connections to other things that nonscientists might be curious about or be attentive to.

A cadre of science journalists have emerged with careers in this science transliteration and transformation, and they are mostly very good at it. The science we get to hear about is the science that they hear about, and that they think we might care about. Since they are embedded in organizations and institutions, a simmering collective mixes what journalists, their editors, and their owners think is important. Mingling with scientists, they size up what scientists like to think is important. And they watch like a hawk their readership, since a journalists' success is the number of people they reach.

As the public, we depend on these science journalists for our fix of science, and while we can shop around a bit, in general they don't come with a focus on every specific area. We depend on their eclectic taste for interesting stories to filter from the vast web of research available.

While it might feel like there is a lot of science reporting out in the wild, the number of professional science journalists is far, far smaller than the number of scientists. The great rise in our general interest in science led in the UK to a doubling of the number of dedicated science journalists over the fifteen years from 1990, reaching just under a hundred people as newspapers and the BBC responded to the demand for depth. The UK is more protected than most countries, with a strong nationally funded public broadcasting service. Since 2000 though, the crisis in funding print media has seen paring back of such specialists everywhere, particularly in the United States, where the focus on volumes of readers to generate advertising revenue tilts against specialized reporting. Even including part-time contributors to columns, blogs, and nonmainstream media, there is perhaps one commentator for every ten thousand scientists, with far fewer in many regions of the world. Like foraging bees, they bring news of distant nectar back to our hive.

WHERE DO JOURNALISTS FIND THEIR STORIES?

You wake in the morning, the patter of news drifting over your bed. At breakfast you scan a few of your news websites, and then in the car you listen to more on the radio. But imagine your day job is creating this stream of news, and more specifically the science content for it. What is your day like?

Let's take a science journalist at a national newspaper first. They might have four stories to file in a day, a few that they started the previous day, but some that they'll have an hour or two on at most. They check their favorite password-protected website, which has over two hundred press releases aimed at them, and maybe twenty by e-mail from press office people they know personally who give them a heads-up on stories up their street. In sixty minutes they have to scan all this, as well as the big stories they were fed like everyone else on mass media while coming into work. They read everything at skim speed, depth flicked over, waiting for inspiration to pounce on suitable nuggets.

Now they have an editorial meeting. More often than in the past, science journalists can influence major stories that contain a seam of science. But the battle at such meetings, which decide the content of the main newsprint pages, is whose story will play to a larger number of readers. While our science journalist might be pitching a few possible stories, more often they find themselves trying to fight against running a science-related scare story. It might be another failed cure for cancer, a food additive bad for your health, or a medical scandal. Such stories are adored by editors because readers (us) really do care about them. But typically they are built up from a nugget of real science, elevated through a translucent stack of shimmering assumptions, and pealed loudly with unjustified conclusions. Responsible journalism can often be suppressing stories that by their repetition lull the public into expecting a constant megaphonic patter.

Science journalists scan the most recent issues of journals that scientists prize most: *Nature*, *Science*, the *Lancet*, and maybe one or two specialist journals that they pick stories from occasionally. Our writer has no time to scan more widely and typically reads the commentary and associated press releases rather than the actual results. They have to assume that everything that is really important will be published in such broad-spectrum science journals, or be commented on there. Since this is the main route through which science meets the outside world, all scientists crave having their research published in these journals. This sparks the fierce competition to get research printed in their limited page space. Journalists and other scientists want to be able to read all the science in each issue, so these publications mustn't grow in pages—part of their value is in staying thin (like the struggle of constant dieting).

In this public ecosystem, the predator/explorers who look for low-hanging fruit thrive best. To look at the major influences providing selection pressures, I will now divide these into *press release patter*, *hooks*, *money*, *images*, *controversy*, and *triggers*, and consider each in turn.

FLUFFY MARKETING

The days that *Nature* comes out are a sprint. Every research article will have press releases alongside, prepared a week before both by the journal itself and by the authors, universities, institutes, or companies associated with the research. Why are these press releases needed? Signposting is key—these embellishments are a necessary accompaniment because the science reported in a top research paper has to be surprising, new, or extremely important to many scientists. The limits on the size of each journal issue mean that the pressure to produce ultracondensed summaries of the science is extreme. It is not uncommon for colleagues and myself to argue (exhaustingly) over every single word, aiming for complete clarity in as few phrases as possible. This does not make for text that is easily speed-read by a science journalist, or even an experienced scientist from a completely different field. Each article makes a brief case for the implications of their results. While vital for the journalist, they attempt to think beyond bare conclusions to build a story that will interest their readers. This is what press releases try to do for them, feeding them hooks and angles.

A press release (PR) is a text nugget showcasing the implications of an advance, and trying to catch attention (like the fragrance of flowers). It will have a title that grabs you like a two-inch-high headline. Written in a characteristic pattern of less than fourteen cascading sentences designed to flow at speed (but rather less compelling than a sonnet), it gives the gist of a research paper, something like a Twitter announcement. To offer journalists ideas for human interest angles, it will have quotes from the scientists all ready to use. Resembling an advertising slot for the research, press releases cannot contain anything requiring thoughtful untangling, and they have to slide unstoppably into the mind of their hosts. They are targeted memes aimed to produce fascination at first sight.

Writing such summaries is an art, behind each being a dedicated PR author. Their job is to crystallize ideas far outside their own expertise, together with the help (or hindrance) of scientists who have been trained instead to deliver accuracy. It is indeed a place of tension in translation. Many scientists are extremely unhappy with the result of dramatically compressing and leavening their already highly compacted work. But just as friends are often better at describing our strengths to others than we are, PR can do a much better job than scientists. Because the costs of translating science are so great, and the need for compression into pithy sound bites is so pressing, there is no mainstream alternative to the PR system.

Press releases are crucial in flagging up science for exposure. Even if our journalist reads each in thirty seconds, that means an hour goes by just reading all that they might be interested in. From this, they identify perhaps their personal top five to go with. They will be aware that they mustn't miss something that will make a big splash in competing newspapers. They will look for ways to connect to bigger stories than simply the one they were fed, a better way to help readers dive in quickly. They might already have sketched out their angle that differs from the triumphalism of a selected press release.

This vast compression of the depth and subtlety of emerging science elicits different responses from scientists. It is easy to see that parts must get lost in this compression, especially uncertainty. However only by compression is there any chance to scan the vast landscape of active science from a ledge further back and higher up. Every scientist struggles to read or even keep up with a tiny fraction of the science directly in their own research field, never mind over the grand landscape. This discouraging deluge is one of the great unsolved problems of scientific life, which accessibility through the Internet has only exacerbated. Finding specific pieces of information on the Internet is much easier than past hunting methods, but getting sensible précis that summarize evolving areas and encircling information of interest remains tough. Intelligent filtering is only one part of the challenge, since ordering the areas of information, linking them together, highlighting possible areas of importance, and providing the right amount to be digested by a reader are all crucial further components to the problem. This is one of the big challenges that future science progress depends on addressing: the automatic collating, sifting, and ordering of science information. In one vision of the future, artificial intelligences succeed in seeing the whole web of science, all at once, that we cannot.

Further complicating the issue is the narrow bottleneck of selection for this compressed science, involving few people and few criteria. This transforms any selection into a political act. Press releases are embedded inside a critical node of the scientific ecosystem, making the question "who chooses which science we hear about" even more vital. We depend on press releases to keep the science we hear about broadly distributed and interesting, but it is more suited to only certain stories. Journalists know readers very well: how long we will read for, what we want to hear that will keep our attention, and how images are important, as well as crucial connections to other human-faceted stories. The formulistic construction of media pieces both satisfies us and programs us to expect more of the same. Nontraditional media, even if widely accessible such as blogging, do not (yet) have a satisfying inner rhetoric and are not driving us away from the PR writing and structuring style.

Beyond the press releases, which sketch what to highlight from the vast landscape of stories that lie behind, are wire stories put out by press agencies such as Associated Press (AP) who even supply full textual copy to slot directly into news vehicles. These have become essential for smaller media publications that lack resources for specialist science reporting. Instead, a journalist with the brief to skim the daily wire reports picks off a story that seems interesting to them and crucially, to their readership. The effect is that any science that makes it to a wire story is then very widely reported in online publications and local papers. Their essence is something easily appreciated and easily illustrated, because the agencies need to recoup their reporting costs by getting onto as many sites as possible. Only particular sorts of science stories surge through the media networks, relayed onward, echoing whispers off the media canyon walls.

Occasionally journalists do have time to meet scientists, but then normally from a small subset emerging into notoriety. Even a diligent scientist in a particular field has a difficult time reconstructing overall progress, and the few journalistic depictions such as James Gleick's *Chaos* concentrate on nascent fields with few simultaneous actors. In most fields with casts of thousands this is horrifyingly impractical, a chattering that far exceeds the largest wedding party.

So journalists mainly get their stories by filtering a rich conveyor of morsels, dangled in front of them for a brief time before whipping out of view. Because of the appetite for science, there is always a ceaseless infinity of stories to be gorged. Which ones do journalists bite?

WHAT MAKES GOOD COPY?

Given the polish and expertise applied to each press release, won't they all make good science stories? Personally I believe that almost *all* science can be satisfyingly translated, given time and resources. As philosophers from antiquity onward have noted, if you can't explain your thoughts to a six-year-old, you don't really understand them. But the question that really determines what we hear about is will it make compelling copy that rises above the other ideas around. In the landscape of competing information, Darwinian-type selection is strong (a key aspect of memes). Given the small amount of time we devote to feeding our interest in science, what is crucial is how sharp and glittering the hook at the start of the story can be.

Whereas most scientists are adding pieces to a puzzle at the end of a long chain of accumulated facts built over years, most media science stories start each time from the same base knowledge of their assumed readership. Any shortcut that avoids initial chains of explanations is preferred. Stories that relentlessly circulate the global currents of the media, such as climate change, tend to be added into this implicit base knowledge. Stories in these areas have a head start since it is easy to jump straight into them. Stories off message for each society have a much harder battle, though they bring freshness and novelty.

Science journalists digest PR to spot the fastest way into a story, and what is closest to their readers' experiences that will connect them to the message. Depictions can take well-traveled routes to their readers, through familiar images from the natural world such as swirling nebulae, shimmering butterflies, or clouds whirling around our planet. These function as icons, shorthand, at-a-glance representations that load up a set of science assumptions that can be used in the following script. Well-used similes are similarly helpful because they trigger-load into mind our prior knowledge. One of the real challenges of a science popularizer is dealing with a widely diverse audience who do not have any common triggers accessing this shared knowledge.

How the story matches the skills of the journalist becomes one of the strongest filters. Does the hook match the readership? Science with eye-catching implications is always easier, with key words including "sex," "cancer," or "money" impressing heavily. A selectivity that prefers science close to homely experience also selects for science that can sound frivolous, the science of kitchen, street, or bedroom. Such selection bias can then prompt us to question why we fund all this science anyway—it has lost is majestic roots.

Another emphasis used to stress the significance of stories is the cost of the research. If a project is expensive, then it has needed to win the support of many people, who collectively act with power. For journalist and reader, it seems it must then be an important story. The assumption here is that important science is expensive, which is not always true. Wonderful science needs imagination, not money per se. Some science is indeed expensive, but it depends on how aggregated its goals are. For instance, we can view the particle accelerator at CERN as a single experiment or as many smaller experiments that work relatively independently (though they all need colliding particles). A scientist might typically spend a four-year research project there working on the design of wiring in a particular particle detector and how they will change the way different subatomic particles can cause erroneous indications of bizarre events. Despite such fragmentation of subgoals and because of our bias to bestow importance with finance, inevitably reporting of CERN concentrates on an iconic massive $6 billion overall cost. This is another sort of icon, used to bring a set of assumptions to mind.

Science areas that allow aggregation of their tools or goals have thus developed a better elaborated set of assumptions within our societies. They aim to cure cancer. Or discover the mechanism of the universe. Or understand the human genome. Or the weather. Conversely some disciplines that have not gelled around single messages or questions are harder to report. The focus for them is on differentiation from others' related work, and advances come across as detailed and narrowly specialized. Although such groups of scientists can see this difference, they still find it hard to invent branding and overarching slogans. We hear little about understanding "everything that can be done with light," "how foams work," "the mastery of all polymers and plastics," or "how cells have living scaffolding and architectures." One factor that differentiates how scientists pitch their justifications is how easily they can directly relate to our normal lives. Curing cancer seems to need no explanation, though the human race might be in trouble if it was achieved, due to overpopulation. Making electrical conductors transparent doesn't seem so pressing and would have been mystifying to lay readers a century ago, but is crucial for all our current electronic gadgets. In each case of course the science might be equally challenging and interesting, and both lead to important results and constructs.

Information-loaded images on the other hand are less important than scientists might believe. Their diagrams are designed to connect and appeal to scientists. Instead journalists and editors use "trigger images" as another hook for a story and thus have no worry about how peripheral they might be to the actual science.

And they are right—the best pictures are those that compellingly trap your interest.

In the end, we reap the science reporting that we currently deserve. Curiosity can be bored or overwhelmed, and we are insanely attracted to novelty. But we have our innately human inquisitiveness about the world, which good science writing can stimulate in many ways.

WHOSE JOB IS TO GET IT RIGHT?

But who is responsible for the accuracy of a science story? What do we mean by "accurate" anyway? At one end of responses, I personally think that any science reported at all, in any way, is beneficial. A substantial fraction of humanity's thought is devoted to science so reporting stimulates, motivates, and harnesses more of our creativity and awe in the world. Even if the science (or reporting) is actually "wrong," it can still turn positive in the long term, despite short-term consequences (such as global virus scares). Rarely will any reader remember the details of X that they have read, but only that it was something about X that intrigued them. Research shows that despite our worship of the written word, we remember less than 1 percent of the details from books we read last year, except whether we liked them or not. So even wrong science can be a woken scientist.

It depends though on what is wrong. Science that has few emotional overtones is robust to a "more-or-less-right" sort of treatment, since apt mental connections become reinforced on multiple encounters. Such looseness is not how scientists are trained, where details matter and every single report should strive to get every single part as correct as possible. Scientists twitch at journalistic mangling that flattens subtlety in order to present a linear story. This opposite response to my own inclination believes that reporting should be as accurate as research articles.

Such polarized reactions miss the recent rise of direct reporting where participants bypass the filtering of media professionals. Blogs in particular give access to this sort of reportage. Scanning successful science-related blogs, it is quickly apparent that scientists who provide detailed information on their own research rarely appeal to a wide readership. It is only those scientists who have become especially skilled in science journalism techniques, not writing about their own specific research but about others in their disciplines, who manage to appeal strongly. They extract clear and provocative messages and provide both information and the tantalizing enervation of burrowing into real science. This science buzz parallels our satisfaction from media entertainment while feeding our human attraction to puzzles, curiosity, intrigue, mystery, and stories.

WHAT MATTERS?

What details do count then in science journalism? What responsibility do participants in this have? Impressions matter most here. Rather than the specific story, it is more important how readers fit the ideas into their picture of the world, a new colored stroke among the pointillist scenes of science. Does it attract emotionally colored blood-red danger warnings? Is it accreted into the white-hot blaze of technology advances and new gadgets? Does it link the verdant green mysteries of the natural world? Here there emerges a tension between the fidelity of representation in this landscape of science, and the immediacy of its interest to us. Accuracy often yields apathetic attention.

Most of us collect statements of what might harm us or be good for us, in a manner that weighs them on our "worry scales." We have all read endless stories of miracle natural essences or dangerous chemicals in our environments, oscillations on the endless swells of the sea of health. I collect these in a mental cupboard, labeled "If it's really important, it will become really obvious," and downgrade the stated importance of all reported stories in this area. I'm habituated to skimming reports, mostly full of alarmist but selectively quoted statistics. Clearly the authors know these reports will be widely read—they appeal to us because they directly make us consider our own daily lives. But each of us makes a decision about how significant to rate such reporting, and it is hard to change people's mind through new information delivered this way. Like our political leanings, we sift these stories to confirm our prejudices and often feel able to accept or reject them by mistrusting their accuracy.

Journalists also respond to our need for amusement, anecdotes, jokes, and escape. In selectively focusing on such stories, they influence high-impact science journals that pay great attention to which research stories attract large followings and are reported widely. Some fields have blossomed because of this, with their branding within the public imagination underpinning growing scientific narratives. "Bio-mimetics" is one such field, feeding "nature got there first and nudged us to develop a new way of doing X" stories. Such diverting science adds to healthy diversity but risks disconnecting us from how science mostly works.

THE RISE OF SCIENCE BRANDS

Science branding exploits the development of icons to signpost shared knowledge. This allows journalists to avoid some of the preamble that prevents a simple and rapid jump-in point to stories. When an inspired scientist bequeaths a new name on a phenomenon or component (such as a subatomic particle, chemical reaction, or gene fragment), it is capable of eliciting compelling power. Naming is framing—framing terms of discussion, debate, emotional responses, or value. But just as branding in the commercial context can convince you to buy something you didn't initially need, so in the scientific context it can weight value in ways you might not accept. As advertising knows well, there is only so much room for global brands in each market, and science is overwhelmingly global now. There is winnowing and competition between different science brands that plays out in both the public and private spaces of science debate. Good branding associates value to a product, implicitly connecting them and framing perceptions. Science branding works in this same way.

"Genes" are an interesting example of this, because the brand image of this concept is so powerful that it is proving difficult to move outside its power. Breaking up the interminable script of DNA into meaningful instructions is far more complex than first believed. Multiple overlapping DNA fragments, chopped up and reassembled by different cellular machinery, finally elicit the molecules that do things in our cells. However this chopping up and reassembly is far from the simplest idea of a gene.

Science branding has also to keep up with fashion, which is extremely fickle. "Knowing the mind of god" and "The theory of everything" are brands currently attached to particle physics. Yet they have become less powerful with time, attracting an air of liability, perhaps reaching that of a "toxic brand." That the science involved now finds it hard to shake off precisely this layer of values attached to them shows how sticky they are.

One encouraging feature of science branding concerns the connection between science fiction and current research. Sci-fi writers mine science that they hear about in the same way as we do and are similarly intrigued by new developments. But they also evolve more deeply the backgrounds to their plots, going beyond the merely popular current stories. As science fiction has grown, it has started to fill the landscape of science possibilities, and thus become predictive for some science, particularly those where evolving current technologies are married to emerging science research, the fields of constructors. Two effects are interesting. One is that when new science breakthroughs are published, science fiction sometimes got there first, and we have been waiting impatiently for the real science to catch up. It's often disappointing that ideas aren't always immediately realizable. The second is that scientists are now mining science fiction for new science branding, that is made to measure for them. "Optical cloaking," "space elevators," and other terms piggyback on implicit shared ideas in mainstream media, enhancing the impact of their fields. To make high-impact science now you *first* write the science-fiction blockbuster that makes it into movie form. Then your audience is ready-made for your trendy research.

COURTING CONTROVERSY

Controversial stories always play well, because they reliably produce the compelling narratives that we prefer to read. "Reliably" here does not mean that journalists are lazy and respond to their time pressures by cutting as many corners as possible. It refers to what we ourselves have a predisposition to read. Science distorts common sense in much the same way as a postmodern novel with repeated flashbacks, points of view, authorial instability, and revisioned scenes requires more time to engage with and digest. So black-and-white controversy really does help frame debates simply. Controversies of interest to scientists, such as whose theory is a better representation of the world, are hard to make powerful. It works for Newton versus Einstein but doesn't work for detailed subjects we don't yet care about. The controversies that have more inherent power are those to which we feel emotional attachment because the implications affect our lives, such as climate change.

This has led to a style of science reporting on emotionally loaded issues in which journalists seek controversy and the contradictory scientist. The fact that there are different weights of opinions on each side, that consensus can be close by, is not what they are aiming to represent. This often makes scientists very uncomfortable or irritated, because the controversy appears to be idiosyncratically manufactured. This style of story unfortunately allows readers to escape from conclusions. Scientists often believe that a *meaning* of their research should clearly emerge from a story, and lead to action. The journalist is more concerned to have their article read to the end and to be engaging. So scientists actually want political aims (for action) encapsulated in the reporting, while we as readers are resistant to such political agenda (supporting the journalists).

Since we like meaning in stories this enhances opportunities to explain new scientific *technologies* that have potential to change our lives directly. The increasingly direct influence of technology on society has driven a good part of the increasing interest in science. While much of the technology for our practical living was available a century ago, it is the recent rise in technologies that work through our relationships, ourselves, our interactions, and our leisure time that seem more stimulating to science curiosity.

If personal experience were the only route into science, what overall view would we accumulate? We would be aware of much short-term biomedical research, focusing on genetics, epidemiology, nutrition, and infections. But only fragmentarily where these touched on deeper areas of science would we see beyond: the complications of immunology, our symbiosis with bacteria, the development and differentiation of cells, or the mechanisms of walking. Similarly, we are aware of new efficient light emitters, of new drug molecules, of 3-D displays, and of iris scanning. But not of how sheets of electrons have peculiar properties in magnetic fields, or of new chemical reactions that aid rapid formation of polymers, or different ways of connecting information in databases to pinpoint what you seek.

We thus form an impression that the science involved in these technologies emerges fully formed, from the start engaged in technology questions. However this is most frequently not the case. The underpinning research was driven for many previous years by other reasons, which only loosely connect even to the same eventual application area. Our view of constructor science is comprehensively skewed toward directly applicable motivations. This is why scientists still keep control over what science gets done, as we shall see.

WHERE DO RESPONSIBILITIES LIE?

One issue that comes up when scientists and journalists argue is responsibility. Who needs to get a story "right," and whose "right" is right. Despite strong influences on journalists, they are not typically overeager to make a splashy story from nebulous origins. They inhabit a community where reputation is important in the long run, in terms of contacts, approval from peers, and recognition. Journalists find their efforts in editorial meetings are arguing to kill a speculative story without solid foundation, rather than overpush something that they have written. On the other hand journalists do not want to cede control of their writing to the scientists whose work they are covering. They aim for the balance between factual and fantastic, to keep readers hooked into a piece. In this part of the science ecosystem there are no policemen, and the wronged shout in the dark.

That is until something really erroneous catches fire and instead of vanishing into the ether after publication continues to grow a life of its own. Stories like this have clear emotional hooks, such as the UK furor surrounding the measles, mumps, and rubella triple vaccine wrongly suggested by Andrew Wakefield in 1998 to cause frequent and serious side effects. The difficult issue in such an expose is to identify where the public interest lies: in exposing a potential major issue possibly being suppressed, or to support the key societal interest in vaccination. Other stories can emerge where a media-renowned figure makes a pronouncement on an area important to them, but in which they are not expert, such as climate change or nanotechnology. Such public spats don't often have a right answer, leading to fishing expeditions by journalists seeking new meat for further coverage and stoking up as many conflicting views as possible.

In many such cases the mantra of public interest is a screen for very personal interests. Scientists desire the diverging paths of notoriety and a quiet space to pursue their interests. Journalists similarly weigh reputation against the massive story. These pairs of practical outcomes clash with/clutch each other in a symbiotic hold, determining what we hear about, and what we want to fund.

WHAT GOES VIRAL

Some stories stop being read as science and become merely part of news. The difficulties of the Japanese nuclear power plants after the tsunami in 2011 took on this life. Here science and technology form part of the essential backstory and become lodged in our minds more prominently than otherwise. Perhaps only in this way can we form balanced views about the benefits and risks of particular energy strategies. From a series of crises highlighting our dependence on oil from other parts of the world, our stewardship of our own natural resources, and the difficulties of long-term storage of toxic materials we can start to weigh in our own minds what choices we favor.

The desperate demand for rapid new information in such crises leads to a hunt for scientists with authority, with strong views, with immediate sound bites. Such rapid mining of specialist subfields emphasizes the fragmentary nature of our science knowledge: there is no overall view of a community, but different threads in their tangled skein of knowledge, heading off in different directions. To help scientists and journalists bridge this gap, a few organizations have formed to act as clearinghouses, helping promote responsible reporting (such as the Science Media Centre in the UK). They have become respected, often because of their ability to connect the right journalists to the right scientists. Formed to prevent media scaring the public with wildly inaccurate claims, a vague shiver of repression in my bones fears such outlets exist mainly to damp down and spread calm.

Occasionally other science stories leap out of their dedicated frames and rampage across the media of all types. This is when we catch a whiff of some future technology, or bathe in a mystery on matters too complicated to normally pin into our framework. Sequencing the human genome, cloning a sheep, high-temperature superconductivity, cold fusion, discovering planets around a nearby star, faster-than-light neutrinos: all these jump into the mainstream and engage for longer than our usual minute's intrigue. In the last five years, top stories of the year highlighted by the magazine *Scientific American* have split fairly evenly between physical and life sciences, with only half as many in engineering or IT. We don't hear too much about chemistry or math, but we do hear a lot about "big science" (i.e., expensive). Many of these stories have both simplifier and constructor sides to them: looking for particles at CERN required construction of intricate technology, understanding what prehistoric DNA tells us depends on recent genomics, or understanding the human immune system gives new approaches for therapeutic interventions.

Some stories promise to overturn everything we know, marking a revelatory point in time. In fact overturning science knowledge is incredibly rare. The scientific process is pretty good, and with so many scientists, almost everything has already been thought of even if only briefly (also many ideas that are wrong).

Viral stories are distributed unevenly across the scientific landscape. Trawling through the (rather partisan) yearly top-ten science breakthroughs from different sources such as popular science magazines, highest-impact journals, blogs, and newspapers tells a fairly consistent picture. Science on our origins, whether prehistoric or planetary, dominates. Another slew of breakthroughs relate to how microbiology and genomics informs health, whether in cancer or psychology. A third domain is the world we live in, environment and climate, flora and fauna. And finally there is a concentration around the fundamental, particularly in physics. Surprisingly, disciplines such as materials science, engineering, and chemistry do not throw up contagious stories—instead they tend to be showcased through strong links to new technology instead.

These four areas of focus depict the symbiotic relationship between news organizations and their readership, reflecting how science and technology news is now broken up into domains. Since these news outlets are responding to how their audience best find what they are looking for, such domains are self-reinforcing. A selected number of stories are scoured in each area approaching every publication deadline, influencing what we then encounter.

The distinction between disciplines within universities is not to be found in the journalistic landscape. Most lab-based science is reported as leading to technologies, and that is how readers (us) most like to read about them. On the other hand, the most fundamental science is reported in its own right, tending to avoid utilitarian rationales but appealing to our basic interest. So we do not hear very much about fundamental advances in chemistry, such as new forms of electrochemistry, new reaction schemes, or the rise of approaches to synthesis that are "green" in their efficient use of feedstock chemicals.

WHICH MEDIUM IS BEST FOR SCIENCE?

The traditional triumvirate media of print, TV, and radio are no longer our only sources. A big change in science reporting is our willingness to read digital social mediators who were not paid for their work in translating into our language. Their writing is less critically appraised or policed and democratizes but fragments our common store of science chronicles. So far though, social media does not drive science or science reporting.

How many of us read the science that journalists write? One 2011 estimate comes from analysis of internet traffic and suggests that nearly a quarter of the US population look up science stories each month. Another estimate comes from the circulation of science magazines like *New Scientist* (137 thousand), *Scientific American* (491 thousand), *National Geographic* (4.48 million), or *Popular Science* (1.34 million). Print figures are converted into readership using various assumptions depending on where those print copies go (something like a two- to threefold multiplier), so perhaps 3 percent of the population regularly read a science magazine. This is about ten times more than the number of scientists. Interestingly in China and India, where science is growing fastest, popular science magazines have done so badly that overseas joint ventures like *Scientific American* have recently pulled out. Speculation suggests that they lack the crucial talented science journalists so far, or that people see science as a way to make money and not as a leisure activity. Science is seen as far more utilitarian in these new science cultures than in the West. Overall no one medium dominates for all groups of readers, and stories tend to appear in all of them, repeatedly swelling hubbubs echoing each other.

BOOKS

Despite the rise of blogs, traditional book form popularization of science has also grown strongly. A number profoundly affected me (such as Douglas Hofstadter's *Gödel Escher Bach*), and I am certain they spurred me into science, as for many other researchers.

There are some peculiarities to the book genre though. Theorists are particularly favored with biographies, especially in the physical sciences. Various reasons have been advanced: experimentalists are more anonymous since they work in larger teams, and it is often hard to pick out individuals; experimentalists are less often eccentric; theory is closer to the supernatural/spiritual but can be more easily read at a surface level; and experimental wrong turnings are just not interesting in the way that theoretical blind alleys can be. Often the authors of current popular science books are also theorists. Perhaps their work habits are already tied to the keyboard, while experimentalists are distracted by the physical gubbins of stuff they have to tame. And of course, then theorists are more likely to write about theoretical advances.

The science journalists who have specialized in distilling copy from science press releases often do a superb job at expanding this to large-scale scientific narratives. Those books with wide appeal use handpicked personalities entwined with the chase of research to mingle interests and relate something of the real nature of scientific discovery. But selecting only the successful scientists gives us the idea that the pot of gold awaits all who venture forth. Most depicted lives are dropped at the moment after discovery, rather than following how it affected the individuals involved (sometimes negatively). Beware the cult of discoverer.

DOES THE SCIENCE WE HEAR ABOUT INFLUENCE SCIENTISTS?

We have now seen one ecosystem feedback, linking science to the media and then the public. We will later see how this indirectly feeds back into pressures on the science. But it is also interesting to see what scientists themselves like to read from among science journalism, as this also influences their behavior when they judge each other's work.

After years of devouring popular science, I became increasingly leery of being influenced by the background agenda faintly but insistently whispering through each science story. I now avoid anything in my own field, seeking more and more distant disciplines where I can engage my sense of fun and interest without the glimpse of politics and lobbying undermining it. So the very thing that had sparked my interest years ago is now in danger of undermining it. Many scientists echo similar views about being disheartened by the relentless rise of posturing and grandstanding, by the need to make ones work "important" and to *demand* an audience. This is a response to the competitive culture now at the heart of the science ecosystem, but perhaps not so different several hundred years ago when scientists vied for priority in discoveries through their national academies. What has changed is the vast growth in scientists, which has amplified this competition for airtime.

Good science journalism is inspirational, to scientists as well as everyone else. While scientists might have a head start on some of the background, they normally come from a completely different research area. With the sheer volume of science being produced, I can concentrate only on my own field's output, rather than struggle with highlights of another. So science journalism provides a feedback between widely separated parts of science. It suggests new possible ideas for links and stimulates individuals by providing them alternative targets for their research. It often provides me the basis for conversations with a colleague in a field I have read about, kick-starting dialogues that otherwise would stay dormant. Science journalism is thus essential for healthy science too, and internally important to its ecosystem.

Despite this it is not very powerful in driving science. Just because research has been showcased widely in the media does not generally increase my respect for it. Experienced scientists take a skeptical approach to being told that something is "important," and all the conflicts raised in this chapter are part of their internal filter that judges stories they read. Scientists see some of the process that makes stories go viral, and being inside this system lays bare the pressures and rewards.

Twenty years ago as part of a team at Hitachi, I investigated how short pulses of light are absorbed by chunks of material. Our experiments showed how wave-like properties of light are imprinted on the chunk's electrons and that by wave interference we can suck back out all the optical energy within one ten-million-millionth of a second. We patented a number of ways to construct extremely fast optical switches, but Hitachi also wanted us to showcase their successful investments in high-tech research by generating press interest.

They employed a smart press agency, and we scientists were shepherded into a process of science translation, producing background materials and nifty demonstrations using sound wave analogies at a press conference in London. The agency used their contacts in the science media to bring along mainstream science journalists. This jamboree produced many (near-identical) science stories in media outlets including newspapers and magazines around the world, following the message that we had devised.

Adding Hitachi's investment in science to their PR spending ensured lots of media coverage. Although not likely to be the best science of that year, its subliminal story was good—that a large industry doing fundamental research might uncover astonishing technologies (what Hitachi called "pole-star" as opposed to undirected "blue-sky" research). The coverage also likely influenced Hitachi management to invest further in basic science. My discomfort though is this easy translation from science to airtime through garlands of money (here tens of thousands of dollars). In the science world, we argue that influence is not so easily bought as publicity, ideas are separately judged, and that the science dies or persists through its use and merits within the bigger picture. My switching work has been taken up in many ways and its origins forgotten (as with many useful science advances, their goal is to become ubiquitous), but PR was not a strong factor in this process.

What about lay readers of the journalists' stories—what did they take away? Did they uncritically absorb messages they were being spun, and if so which parts? I do believe in the messages crafted: that disruptive technology comes from early-stage science, that wave-like properties of electrons can be technologically important, that Hitachi was associated with leading technologies. But money was able to elevate our science over vastly more being done at the same time, and this was translated uncritically into coverage. No one ever probed deeply into the real implications of what the science was, and how realistic it was to use it (all really interesting and real questions). They were just looking for a good science story; they trusted their contacts that this was one such and went about their craft with efficiency.

THE BOTTLENECK OF DISSEMINATION

The process for bringing science stories to the public sphere is clearly very selective. In particular, the controlling "bottleneck" of a band of press translators and opinion formers (figure 6.1) is crucial in this process. Together with the selective role of journal and academic competition, fashions, and human desires for diversion, the ecosystem thus sifts finely for the gold of a story that goes viral.

This realization should make us a cautious in assuming science is done in the ways we read about. Only certain stories are fed to us, the lowest common denominator of triggers for our interest, generally connected either to immediate worries, entertaining vignettes, futuristic gadgets, or idiosyncratic scientists. A vast fraction of science does not fit into these categories but yet often underpins their

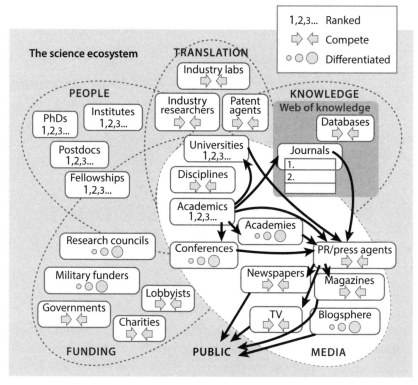

Figure 6.1: Dynamic interactions of actors in the media "scio-sphere" of the science ecosystem. The arrows show the bottleneck at a small number of press influencers.

outcomes. The pressures of journalism restrict attempts to offer different narratives. While the digital realm democratizes our access to science it also fragments it, mirroring the bourgeoning of science papers and conferences that fracture an overall view of science progress. In contrast to the past, it is now very much easier to find the new knowledge emerging from science, but hard to discern any strategy, timescales, or cool-headed evaluation of its prospects. As a result, what we hear in the media about science bears distorted relation to what is going on and how it is done.

This is important when we think about who chooses the science that gets done. Although we might not hear about very much of it, we need to understand how choices are being made for us all, and so that is our next port of call.

WHAT SCIENCE GETS DONE

So far we saw how knowledge moves around the ecosystem of science. But creating that knowledge is expensive, and this limits what science can be done. In this chapter I look at who decides this, and how. Basic research is so risky that only governments can now afford to do it, subjecting it to many diverse influences. We will find levels of support for each government-funded scientist vary enormously, more than fivefold, creating very different science environments in different countries. As might be expected, the battle for resources is the most explicitly competitive part of the science ecosystem, making decision making somewhat risk averse and conservative, because it is coupled to the success criteria that were introduced in the previous chapters. It is as if competing predators and prey on the plains of Africa have sat down and designed a committee process for who gets eaten or fed.

I will first look at how scientists decide what they want to do, and where their resources come from for taking their next steps. Then I will explore the overall resources currently put into science, and the rationales for this investment from societal benefits. Intriguingly no one agrees on how much science should be funded, but they agree not to talk about it. Finally I will consider how decisions are made for which ways science funding is spent, where not only scientists come into direct competition, but also the institutions in which they are embedded as well as the different funders.

RESOURCING SCIENCE

HOW SCIENTISTS PLAN

Most research follows on directly from previous science. Recent results suggest the next most intriguing direction to proceed in, along visions of particular researchers. More rarely, periods of intense reflection suggest an unusually radical direction for a scientist to pursue. Such evolution and contemplation can combine to open new directions by bringing in pieces from other lines of research. However what actually becomes the next focus depends most highly on resources. All scientists dream of something they would like to champion as leader, but the reality is a succession of obstacles. Science is hard.

While scientists will say that they pursue research that fascinates them, I already showed a range of influences exerted on them. What they choose to do next in their research also responds to these pressures, whether to gain interest from colleagues, show potential for high-impact publications, or enhance opportunities for gaining substantial funding. Many of these influences coalign so that problems of major societal importance are tackled. But rarely is it so obvious from the start how a piece of science might impact any key problem—this emerges with hindsight from the results, but beforehand other factors are at work.

The usual direction in a research program is toward increasing complexity. Each scientist works within their core subfield, hunting for the next thing to advance. But all the simple things have generally been done, and so the only direction forward is to build on prior foundations, necessarily creating more sophisticated schemes. This is typical constructor science, although maybe in the service of simplifier science. What emerges is a picture of greater and greater detail, with more and more finely articulated aspects. Sometimes though, almost miraculously, a scientist is able to go the other way, to take a simpler route ahead because their ideas and others have opened up a new space to explore. Instead of slower progress through denser thickets in the web of science, they have come across a large clearing, with views out over the wider landscape. While rare, these are special moments in a scientist's life, when their excitement at seeing the vista unfold is rewarded with exhilaration at glimpsing further connections then they had dared dream were possible.

In biochemistry the next step on the path of understanding how a cell works might be by unraveling of mechanisms controlling a particular gene, understanding all the factors that regulate whether it is active in the cell and at what times, and what else it interacts with inside the cell and the genome. How to select which target gene, or what facet of its operation, is what makes a particular scientist special. It is the *selection* of suitable problems, not just the ability to solve them, which distinguishes different capacities for research. What counts here is "having a hunch," "following one's nose," or "sniffing out a good area," all attributes where the programmatic and logical sides of science are less overt. Even though research evolves from making some initial choices, it is setting the overall direction that controls what unforeseen but interesting aspects might be encountered.

In explorer-led (simplifier) science, the only direction is supposed to be onward toward understanding the mystery—but always there are choices. Scientists in grand challenge areas may claim that they are trying to "cure cancer," "find the quantum basis of gravity," or "understand consciousness," but none of these are specific enough to help decide exactly what to do next. In particle physics, what is the next experiment that a researcher should join in? Normally such scientists have been working in international teams for many years, distilling into them the new directions for their subfield, and lobbying for development of mega-expensive new experiments. Given the five- to twenty-year lead time on securing funding, planning, and constructing massive facilities, decisions cannot be impulsive—they carry a long-term commitment, like a marriage. Often those they trust have emerged from the same or connected research groups, part of the same family of loyalty. By contrast, theorists can often be more impetuous in choosing where to focus their efforts. Generally large amounts of money are not needed for their work, but time is their most precious resource, as well as the theory box of tricks they have mastered so far in their career. So what they decide to do is a more impulsive affair. Scientists thus plan on very different timescales.

SCALES OF SCIENCE ENDEAVOR

The most crucial resource needed for modern science is money. Nuggets of interesting science still flourish without extensive financial backing, mostly in gentle, less-competitive backwaters and necessarily operating at human scales. A favorite of mine studied the ways that a piece of paper falls to the floor, identifying what gives it the swaying flutter we see (which actually helps now develop aerodynamics at insect scales). But even then, next steps might be to simulate this with the air eddies and sheet bending needed, and already this requires intense computational power. Another more recent paper studied the way that droplets of water condense on spider silk in webs and showed that the thread has special properties that allow it to wet and wick water from the air. Again, alongside more simple apparatus, expensive electron microscopes were needed to image on the smallest scales.

Some pieces of research need vast resources, whether in equipment, numbers of people, deep data crunching, or research span. Researchers aren't specifically drawn to this—more often their peer group subfield has taught them the currently accepted way to approach resource issues. It is essential they adopt these approaches to be successful, since it frames the views of scientists and funders controlling the resources they need. Scientists must choose between joining a larger team, going out on their own, or bringing a team together under their own leadership. Previous experiences and the culture of their subfield set what they are most likely to do. If they want to amass a team, they have to first convince others of why their idea is worthwhile. This is crucial when applying for funds and happens even before funding applications are submitted—people first have to donate their time and prestige to a new activity. Building consensus is thus essential even before the science is done that will produce consensus.

Without a magical patron bestowing full freedom and funds, scientists have to seek money from someone. This money comes in several ways: from all taxpayers via our governments, through our personal donations to medical or health-based charities we want to see conquer afflictions, or through our purchase of products that we want now or our appetite for future products we would aspire to have, which encourages new development. Each of these funding sources brings its own range of pressures and possibilities for the researcher, that shape what science is actually funded.

WHERE DOES MONEY FOR SCIENCE COME FROM?

By the twenty-first century, we have a diverse range of organizations prepared to fund scientific research who have been convinced historically that this is a good investment (figure 7.1). The primary source of research funding is from governments, particularly in the physical

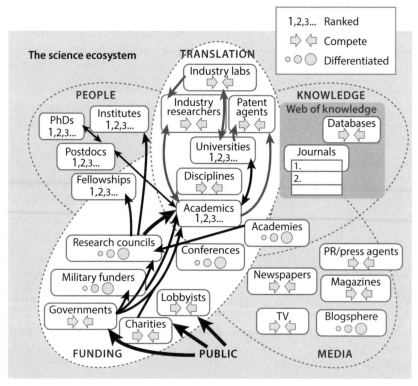

Figure 7.1: Scio-spheres of funding and industrial translation of research, all originating from the public purse.

and mathematical sciences. In other words, the taxpayer has been convinced that science is a good thing for societies to invest in. But it is largely true that the demand for science is not created directly by taxpayers themselves (except perhaps in demands for cures for illness). Instead governments judge it to be an instrumental tool that helps them achieve higher goals to satisfy their societies, such as robust, satisfying work; safe, secure environments; or enhancement of choices for their populace.

Funding comes through one or more government ministries (closer) or "research councils" (further), at varying distances from direct political and civil service influence. These ministries have to fight for science resource at each budget. Science is thus necessarily politicized despite the researchers' desire to disentangle themselves from this mesh. Previously static and closeted consensus about how much resource to confer into science has in recent decades been exposed to increasing demand from scientists and their institutions. Both scientists and politicians have apparently accepted that science is a tool that delivers results. Throughout the world, from the United States through Europe and across Asia, the Vannevar Bush postwar compact posited science as a way to advance society. Thus a broader range of motives has formed around the *utility* of science, and scientists have now clear interest in demonstrating this in order to access more funds.

In some countries, significant fractions of the science budget originate from the military side of government. Around 15 percent of R&D spend in the UK comes via the military compared to under 5 percent in Germany, while in the United States this is over 30 percent and in Russia over 40 percent. Data for Israel and China are not easily available but are likely similarly significant. While often more focused on constructor rather than simplifier research, defense funding in the United States (itself massive compared to elsewhere) contributes to much breakthrough science. Technology has split modern warfare with developed countries emphasizing innovations in order to protect soldiers' lives, while small armed nongovernmental fighters exploit it in limited ways and acquire what is already available. This funding for military advantage is nothing new—it was one of the first formal funding routes for science, from investigations of more stable gunpowders to new navigable routes for accessing foreign wealth.

Many of the inventions we use in daily life came initially from military investment. The laser and many of its technologies emerged from massive military investment in the United States after WWII but now underpin how we watch DVD movies, chat on the move, or weld back detached retinas. Would these have happened without military investment? The interlinked nature of science suggests that indeed this research was part of a developing framework, and would not have been ignored. But the speed with which it developed depended on the scale of that investment. Military floods money into particular parts of science, in the hope of a rapid gain. Without this, most advances would be shared among all nations in the world as the ecosystem of knowledge is profoundly open, and military advantage would be lost. Current military focus on autonomous, bio-, micro-, and nanoweaponry is a new battleground in which such temporary advantage is sought.

Much basic science in the United States is actually funded under such aspirations. Accepted there as part of the science ecosystem, the army, navy, or air force research labs fund innovative ideas. The Office of Naval Research in the United States claims that more than fifty researchers have won a Nobel Prize under their funding. Another military funder, DARPA, is known for its engagement with fundamental science that can potentially lead to novel technologies in coming decades such as quantum computing, while their proactive approach and test beds did lead to early implementations of the Internet. What in many countries would be seen as the domain of science funding agencies has in the United States been siphoned into military rationales, however most scientists themselves blur the distinction. To them, this is just how the science ecosystem is organized, in which they have no control, and they are forced to "play the game." Outside the United States, military routes for funding science are much less important in terms of producing scientific breakthroughs. But in the United States, the Pentagon virtually doubles the total government spending on R&D, with a substantial fraction for constructor research in universities. This is despite studies that show investment in nonmilitary-sponsored research is better for the economy. Recently, the US Department of Defense was criticized for refocusing on short-term science in a report complaining that "in the present program, evolutionary advances are the norm, and revolutions are less likely to be fostered than they should be."

INDUSTRIAL FUNDING

Besides funding from taxpayers by governments, companies play a strong role in developing new ideas—they stand to benefit (figure 7.2). Particularly in high-tech products, but also across all industries, innovation is essential to avoid the competition making your product or services archaic and irrelevant. This Darwinian evolution is a significant spur to developing new technologies through research. The major difficulty for all industrial investors of science is where to focus your limited resources, and at how early a stage an idea might be appropriate to back. Should companies do basic research, applied research, or technology development?

In the second half of the twentieth century, large companies prided themselves on developing the new ideas for their products completely in-house. The belief was that patent protection and internal know-how gave sufficient defenses to allow them to reap the rewards. Drugs developed in a company might produce massive profits for many years. New designs of semiconductor computing chips would repay over many generations of products. However a number of problems emerged from this experience. First, the risks turned out to be high. Despite hiring the smartest people to attempt to define what research to do, many programs turned out to fail. This

Figure 7.2: Powerhouses of science. *From left*: former Bell Labs, where the transistor was invented (New Jersey), former DARPA head office, which funded the precursor to the Internet (Washington, DC), and Wellcome Trust, which funds biomedical research (London). Image credits: L Aberle/Wikimedia; Coolcaesar/Wikipedia.

says something about our inability to make science programmatic—there are too many unknowns that define whether an idea can work or not. The approach says something about our myth of a single continuous path linking scientific ideas toward practice and thence to exploitation. Many unforeseen issues emerge along the way that limit what might work practically and be useful in technology by the time it emerges—ideas from unknown competitors might easily work out better in the end.

Second, companies with a successful range of products based on one technology creating large profits find it difficult to support developments of another technology that cannibalize this product range. Emblematic among these is Kodak, who internally developed and patented some of the first electronic cameras based on semiconductor chips, but didn't want to kill off their enormously profitable business in photographic film. As a result they never committed to the new technology, and others such as Canon, who did, eventually put them out of business (who in their turn are now being put out of business by mobile phones). Commonly in recent years it is small innovative companies that develop such disruptive technologies, and then are bought up by a larger company. Or a larger company might invest with arm's length control in such an innovative internal research unit. Companies thus try to outsource their R&D.

Thirdly it is not clear if it pays to be the first to develop a new science idea into a technology. It costs a lot of money, there are many false paths where insuperable problems emerge, and only technologies can be patented, rather than new science. Science knowledge diffuses around the community, and if it appears promising then other companies rapidly find alternative (unpatented) ways to deliver a similar result. So the originator has only a limited time to be in front. Often it is best to be second in developing a new technology, so that someone else does all the expensive exploratory bit and developing the market, and one instead concentrates on rapidly learning from them to develop more cost-effective and higher-performance adaptations. For this, a highly skilled research team is needed who look outward at what other people are doing, not hermetically sealed off internally to protect its work. And since science often develops profoundly at its boundaries, looking outside is a much better strategy that spurs new ideas.

It is these difficulties that have supported the rise of government-backed research—it derisks the early stages that otherwise would not get done despite their promise for society. It makes available new science to be openly exploited by all companies. As a result, companies now look to coinvest in science all around the planet that might be of use to them. In return for this investment they get some rights to the results, although these rights differ widely depending on the cultures at work. Universities used to give away their results to companies, as this was seen as their mission for society. More recently there has been a sense that companies are getting something without paying for it and thus they do not value or exploit it appropriately, so now universities in some countries try to retain more of the exploitation rights as we discuss later.

With the globalization of economies, companies are less locally minded. Supporting science in their own locality is all very well, but realizing their technological ambitions comes first. They might also be looking for several goals in partnering researchers, besides the obvious concrete output from a project. Engagement of their own internal researchers with others brings new ideas into a company, perhaps from other domains of knowledge, or from the state of the art within a subfield, aiding more rapid diffusion of appropriate science into their orbit. Young university-based scientists who they engage with often acquire more favorable (and certainly more realistic) views of the company, and this is an excellent way to recruit ambitious bright minds. Finally, with globalization also comes associated public perception of companies. Funding science research is a way of branding, introducing links between societal advances and the company, with high-technology seen positively and as forward thinking particularly among the young (see my example of Hitachi in the last chapter). It may be used to offset perceptions that a company is exploiting the public, for instance in cosmetics, food and drink industries, oil extraction, or nuclear power. Academic research investments from these industries send a message that they are not one-sided, that they seek the truth, and that they are also guided by the ideals of science, not only commercial expediency.

As a result, however, companies intensify competitive aspects of research, because instead of talking to local researchers down the road, they might seek out and rent the very best minds half a continent away. Funding from industry can thus become more focused in elite research organizations, which in turn have developed more and more sophisticated ways to woo these industry funders. This raises unease in many quarters of science, entangling commercial return with scientific truth.

INVESTING IN FUTURE PROMISE?

Emerging sciences do not always have a product or a market yet. Why would I want to buy a personal genomics product right now, because it's not yet so obvious what I gain. Without a company interested, how can I get the funds to develop science that might be important both societally and commercially? These needs are filled by investors who believe that investing in science might bring enormous profits. Ranging from venture capital companies to business angels, these finance companies are looking to see what can be extracted as quickly as possible from any investment. They have a paramount need to identify science that realistically can be exploited, and then help to bring it about. Once again, the ill-defined progression of science makes their job extremely hard, with the result that over 90 percent of such ventures fail within a few years. The few that succeed can be wildly successful, luring further rounds of investment, such as ARM in Cambridge, which developed almost all the chips powering our mobile gadgets.

The main problem for such investment in science is not money, but time. Science is inherently limited by the working out of ideas, and throwing more money at a problem is not always a recipe for working things out faster. Only few ideas make it through the barricades protecting the slabs of money entrusted to venture capital firms to use wisely. Slow science that takes longer fails at one of the first barriers, and practical proofs are more important than cool ideas. Not all scientists are so interested in the distractions and burdens that accompany turning ideas into products, while many great pieces of useful science come from ideas that do not work as envisaged but instead lead hazily into new directions. One of the greatest difficulties with doing science in the service of a company goal is that deflections from the core aim are not acceptable, leading to conflicts between the open-endedness of science and the essential aim of companies to make enough money to survive and grow. My own experience though shows this goal-focused science research is not inevitably technology development, but still generates new ideas and knowledge.

In general the science ecosystem is not strongly dependent on developments funded by such investors. However the activity itself is seen by governments as a crucial part of the rationale for funding science. For this reason, scientists face increasing pressure to become industrially involved, which can actually lead to a decrease in the amount of science they do. From a taxpayer point of view, the occasional jackpots dangled in everyone's eyes enhance belief in the role of science in our lives, and its alchemical ability to magic gold from nothing. We will consider the evidence for this later.

TAKING GIFTS

The final supporters funding research are single-issue charities and philanthropic foundations that see science through a utilitarian prism. Can studying this area "cure cancer," eliminate AIDS, satisfy third-world water needs, or a host of other goals? Depending on the attitudes of their management, and their donors, such foundations can think in the long term rather than just ameliorating symptoms now (though they can also be hijacked by companies seeking to promote their own solutions). Donations come from immediate emotional responses but at an aggregate level deliver a working space that opens out much longer timescales for research, from days to decades. Billionaire philanthropists (like their Victorian progenitors) have an increasing impact in this category, with disproportionate media influence and science leverage.

It is easy to state that science can help all causes, but choosing what science will accelerate the solving of problems is fraught. The best ideas are often unexpectedly remote from the problem they might help, so charity investments are better at connecting already emerging science with places where it might be applied. Rather as military funding, they look to accelerate adoption and diffusion of new science into particular problems. In the medical ecosystem, charitable support is one of the major funding sources for research, and fiercely sought and esteemed. Environmental charities use scientific knowledge to respond appropriately to their challenges and to lobby for governmental activity, so research results increase their power to influence change. Development charities seek to bring attention to needs (and hence new markets), and to develop pathbreaking solutions that can be widely adopted.

Taking funds for research from a charity deepens connections to the moral aspirations of many scientists, seeking to change the world for the better (though researchers taking money from military will actually say the same). Such feedbacks complicate the science ecosystem since rewards become split between internal and external spheres. Useful science may not be exciting science. The effect is to drive fragmenting wedges between actors within the ecosystem, depending on how scientists and others choose to see their roles. The ascendancy of esteem is perhaps still with those at the core of disciplines, who see the web of knowledge as paramount with a halo of benefits centrifugally hurled out.

The need for diversely funding science is thus evident, enabling scientists to take forward their different ideas, whether intellectually or technologically inspired. What I turn to now is how much is currently spent and why.

FUNDING RATIONALES

HOW MUCH MONEY IS SPENT?

The amount of money going into science sounds large in absolute terms (and governments have been persuaded to compete with each other to supply totals that sound increasingly impressive), but such funds are slight in terms of the fraction of the tax we pay toward it, or the much larger amount we fund health care with. Developed countries spend between 1½ percent and 3½ percent of their GDP on science R&D, of which 30 to 40 percent is funded in some way by governments. However, only 0.2 to 0.4 percent of the GDP is directly provided for scientists seeking funding for their research projects, about a tenth of each country's total spending on science R&D (the rest goes on infrastructure of the research environment, and the salaries of academics and administrators). The absolute funds are thus rarely substantial, with about $100 per year for each person in the population being directly invested in science research projects. These numbers have been fairly stable though gently increasing for some time, with governments expressing strong interest in trying to get their industries to spend more on R&D. Politicians, as cheerleaders and funders, have been convinced that research pays. This has then convinced countries like China to massively ramp up spending on science, currently providing less than $10 of project investment per capita, but doubling this every four years.

Another way to look at investments in science is to see how much funding an average academic gets to spend on their research projects. This will conflate wild fluctuations between different disciplines but suitably contrasts the regimens in different nations. Funding available for each business-lodged research scientist in the West of around $200,000 per year varies little (within $50,000). However there are much greater variations in the basic funding for academic science between countries (figure 7.3). I am using here OECD numbers and having to assume a similar fraction of science to non-science academics across countries. On this basis, the UK invests the least in each researcher, around $20,000 per year. This is half that of Spain, Canada, or Austria, while Japan, France, and the United States all reach $80,000 per year per researcher. Most princely is Korea, which is spending over $100,000 per year per researcher, more than five times that in the UK. Studies by the UK government have shown similar conclusions. No wonder that the UK ranks as the most efficient in the world at research impact per dollar invested. Inevitably it is not just that the administration is lean in the UK but that the salaries and investment in equipment is sparer. This fivefold variation is startling, given the lack of evidence of its impact on the science output.

We should also remember that government support is only one aspect of funding science. In the UK, additional support from industry contracts, the EU, and charitable foundations more than doubles the research council investment, and even more for engineers and medical scientists. Similar ratios are seen in many countries, but

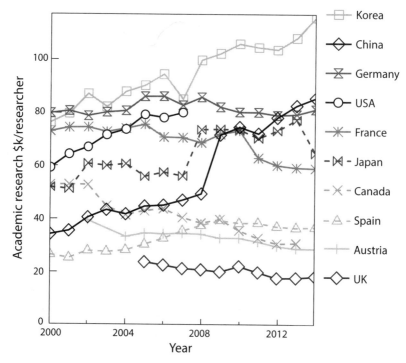

Figure 7.3: Average academic funding per researcher (including all research and infrastructure spending on science) for different OECD countries over the last decade.

there is little clarity on how much nontaxpayer sources support fundamental advances in science (such as Nobels), rather than specific technology projects.

GULPING MORE

Demand for funding in science is driven foremost by the increase in the number of scientists. But even in countries where the number of scientists has settled, the increased competitiveness of the science ecosystem has increased the aspirations of scientists and thus their demands for extra funding to compete. There are now more scientists working in their area internationally, so competition is fiercer and speed is more critical. If they are not funded to work with state-of-the-art equipment provided to competitors, then it is hard to "win" by being smarter. Sometimes, starving researchers of funds can lead to innovative and productive solutions that do even better than buying-in the newest gadget—they can even lead to the next generations of gadgets that everyone wants to have. But research inevitably takes longer if you have to invent tools as well as push on the research itself.

Stepping outside these pressures we might ask what is the appropriate pace at which research should proceed. If we see science leadership as a race in which only one country will win (which makes little sense to me), then each administration needs to decide whether to fund everything or to be selective. This debate is clearly framed around the utilitarian view of science research, with reward directly connected to results. On the other hand, novel science often emerges in unexpected places, and we might assume that funding broadly in many areas is most likely to uncover these nuggets of intellectual treasure.

Research itself is also becoming more expensive. In our previous eras of apprenticeship, young scientists were paid little but sufficient to allow them to focus on their devotion. It is no accident that the modern academic scientist emerged from the ancient monastic traditions of contemplation through pondering the glory of creation. Our current era doesn't demand quite such sacrifice but (as will see in the next chapter) still has difficulties delivering financial security by providing permanent jobs at a young enough age before personal commitments loom large. Professionalizing science research means now salaries are meant to reflect skills, training, and societies' needs. Thus one cause of the increasing cost of research is the cost of scientists themselves.

Another increasing cost has been the development and diffusion of expensive tools throughout the research landscape. With technically simpler problems solved (Newton never claimed expenses for his apples or prisms), scientists study the intricacies of worlds remote from the human scale. Mediating this expansion of their senses are delicate machines that allow them to feel atoms, to video living cells in jittery action, or to feel the faint heat of ancient galactic filaments. Access to such machines is prerequisite for getting involved in modern science, but building them yourself or buying them in are both expensive. If the rationale for research is getting to new science first (harvesting impact, patents, and know-how), well-funded infrastructure is essential.

WHY DOES RESEARCH PAY?

Scientists have a nuanced view about the benefits of research, spanning ecosystem services as well as ecosystem goods, epitomized in "knowledge for knowledge's sake." Most scientists believe spending funds on research is very good for the country they are in. Most governments believe the same thing. But what evidence is there that research spending is good for host countries, rather than a global good? Does government investment really deliver improved quality of life for a country, and in what way?

Science currently follows the "Columbus" model of innovation. Scientists are supported to follow their own ideas, after which discovery and utility will follow. This leave-it-to-me attitude is very deep rooted and very convenient for scientists, but not clearly evidenced. The problem for reaping societal rewards is that times between fundamental discovery and exploitation can be very long. In such cases it is less likely that the original country where the first breakthrough was made will be the sole beneficiary. Science as we have seen is wonderfully global, so that knowledge flows wherever there are researchers who are interested and can accept each package of discovery.

Much has been made of even more local economic stimuli from research. In some places universities are viewed as engines of economic development in a city. While there do exist a few isolated examples such as Silicon Valley in California or Silicon Fen around Cambridge in Britain, attempts to explicitly engineer such links by encouraging entrepreneurial activities and high-tech start-ups have not performed as hoped. Recent studies find very little causal link between high investments in university research and growth of and employment in cities in the United States. The vast number of university research parks on the fringes of cities around the world have generally made minimal impact on local employment, and taken up valuable tax funds as well as community land and vitality that could have been used for more effective strategies. But the myth is very powerful and has been widely encouraged (figure 7.4).

Figure 7.4: Innovation hubs. *Left*: Silicon Valley (San Jose, CA), *Right*: Shenzhen (China). Image credit: Coolcaesar/Wikipedia.

Looking at the science spending in different parts of the world, it is startling how great the disparities are between rich and poor. While slowly changing, the spending in Africa is destined to remain small for many decades. Scientists in the West often work on problems of great relevance for underdeveloped and environmentally troubled societies across this continent. In so doing they are practicing a belief in "trickle-down science"—science supposed to directly benefit a group is done by a completely different group. Rather like trickle-down economics of the 1980s, which assumed the spending of the wealthy would directly help the poor (it doesn't appear to), science assumes that this spending will help the societies that don't have much science themselves. A big problem however is that scientists solve a problem that turns out not to be the key aspect holding back development, for instance, water purification solutions that have to be suited to the climate, environment, culture, and economics of equatorial Africa—and often they are not. Stimulating locally active scientists embedded in these societies is an urgent priority, but as yet is not working. A growing realization of the benefits in longer-term twinning of first-world and developing-country scientists is emerging, requiring a greater fraction of science spending outside solely national domains.

Most important by far from the science investment standpoint is the human resource that emerges—scientists trained to think in their particular analytic or problem-solving mode that is exceptionally useful in all walks of life, from businesses to government, from economics to charities. For a country to emphasize this rationale for funding science, they might want to know how much training is useful, and what fraction is optimal to become the next generation of teachers. In most countries tracking where researchers end up has only just begun.

TECHNOLOGY TRANSFER

In an effort to speed up and clarify this awkwardly nebulous relationship between science investment and economic growth, the last decades have seen swarms of technology transfer offices formed around universities. Like oil companies, these are supposed to extract the valuable assets of science as they are generated at the work face and spin a cocoon of gold around them. Unfortunately the evidence to date shows that most such activities lose money. I particularly like the quote in a 2005 study by Lester and others that "the best form of technology transfer is the moving van that transports the PhD from his or her university laboratory to a new job in industry." In the United States over 70 percent of tech transfer offices lose money, with only 5 percent making a substantial profit (most from licensing new drugs). In the UK, the annual return from licensing IP from universities was only £58 million by the mid-2000s, less than 1 percent of the total research spending. The explicit directly created jobs in the UK are far too few to directly impact employment.

A clear difficulty for effective technology transfer is that industry cannot passively suck up new science results, but instead has to expend a lot of effort in making use of them. They need a threshold of their own internal research efforts to support the tacit knowledge and "absorptive capacity" required to identify and assimilate potentially exploitable advances. The demands on industry mean that making good science is not enough for a country—competent and advanced industries are needed too. The incentive for companies to support their internal research effort is in conflict with the need to generate shareholder value in the short term, because of the frequent long time lag (over twenty years in biotechnologies) between initial discoveries and commercial returns. This is why the "linear model" of innovation in which new ideas and results lead to company growth has been astonishingly hard to provide evidence for. Especially the claim that science success in a country can directly lead to wealth creation in the same country through exploitation of the results is lacking in robust statistical support.

The different goals of business lead to a great contrast between academic research that wants to shout about new results as widely as possible, and industrial research in which as little of the new science knowledge as possible is shared. Scientists want to write, industrialists want to read. This has evolved into a stable paradigm, because of the public good from science. Societies benefit if the fruits of science are well disseminated, so they encourage scientists who publish new results, by conveying esteem on them through prizes, keynote talks, and funding. This is also why academics in the university ecosystem are group selected for their prioritizing of esteem over salary. Otherwise they would have a stronger interest in hiding their best work, and gaining financial advantage through secretive industrial interactions. As we will see in chapter 9, inflation in the currency of esteem and the increasing veneration of wealth in our societies are less supportive of this group selection (for instance encouraging many more academics to found spin-outs), pressurizing science culture to the potential detriment of science for society.

Why is it so hard to show that publicly funded science helps a country? For any new product or service, a haze obscures what emerged from prior research compared to the many other inputs. In the United States, 75 percent of patents cite publicly funded research. But it is very unclear how many of these research papers were the key contributor to the invention, rather than just part of the background landscape. One study has shown that investing $125 million in drug research in 1974 would have yielded on average only one new useful drug in 1993 (after twenty years). Quantifying these advances to society is also difficult. Besides company profits, changes to society are not universally beneficial: preventing early deaths increases productivity, while preventing post-retirement deaths increases the pressures on pension funds and increasingly stresses the economic balance between rich and poor, the young and old.

This considers only the direct return on investment in science projects, but there is another major way that science helps a country. System-wide knowledge of science enhances a society's capabilities. Advances in creating new instruments or methods can indirectly help many companies. Trained researchers emerging from research and moving into not only industry, but also civil service and other organizations, improves their ability to solve problems. Networks of people are broadened in span, bringing new creative possibilities. This research "spillover" is based on people so can indeed have local as well as global effects.

Local spillover effects have been studied in the United States by looking at how often patents refer to papers from local universities compared to those in different regions—this is at least double what is expected from chance. But again, we don't know how often these are useful patents that later generate new profits, compared to those arising from preexisting collaborations or spin-offs. In fact, despite the rise in patenting by entrepreneurially optimistic universities, research has shown that these are being cited less and less compared to other patents from industry. A recent study from the Institute for Fiscal Studies in the UK provides evidence that funding research actually helps the countries that do not *yet* make large profits in any particular industry, aiding their catch-up. The further behind they are, the better investment is research to help them catch up. Thus the United States (at the front of many industries) actually helps the rest of the world more than itself when funding science.

Since the claims for direct and immediate (within ten years) effects of science on technology developments in industry do not appear to be so obviously striking, why do governments listen? There are clearly very many lobby groups providing strong anecdotal stories (such as Silicon Valley, or Boston), which are more compelling than any statistics. Because funding science fills a market failure to deliver a common good of society, we could say that all science is an ecosystem service. In that case, the benefit is for the children of our societies, and their children. These are our future populations that governments in the past decades have been rather bad at standing up for, instead transferring debts onto them while supporting the ageing generations alive now. Governments also have at best five-year horizons, within which no direct benefits will justify the science investments. So the case for science funding is better founded in a compact between the public and scientists, where the government acts as middleman.

CHOOSING DIRECTIONS

HOW TO SPEND WISELY?

So you are in charge of spending a billion dollars on science. How will you do it? What are you looking for? How to be wise? While scientists universally complain that not enough money comes to their own research, the truth is that deciding how to invest in all the myriad different possible opportunities is dazzling and mind-numbing. The merest hint that you have some money to fund research triggers a cascade of deserving research opportunities to your door. Demand never seems to be sated.

A government-funded research agency has to justify to the finance ministry how they spent the funds that they fought for. Many promises have been made, and the agency is judged on turning taxes into "results" through a sorcery that transmutes research funding into science. If they succeed convincingly, their credibility is enhanced and their promises more believable next time round, while at the same time their foundational support of the scientific community is realized. Failure harms the society and the science base. More concretely, the strength of a funding agency is judged on its importance, nowadays measured by total funding and the percentage increase in funding. Importance is also judged through their visibility within society, if the agency is known to be the backer of science breakthroughs, or as the engine for developing new companies and jobs.

At repeated focus groups filled with members of the public, I have heard the surprise that scientists are allowed to choose whatever they want to work on. Why not just tell them what is important for the country and insist they focus on this? Why not hire them for a particular role in an institute that aims to solve a particular problem? Indeed, it happens (sometimes) in industry that scientists have clear roles and their projects are generated by others and assigned to them. But most generally, it is not found to be a very efficient way of doing research. Even problem-focused research can require a breadth of exploration that consciously directed science may not impact. Even more troubling is identifying what pieces of science would help a country most. Who would make such decisions?

In some countries (now typically in Asia), the research agency is staffed by experts who are given authority to dictate. They decide how many chemists their country needs, of what sort, and what teams they should work in. They decide to build a new institute focused on photomedical therapies, or developing missions to mine Mars. Beyond the assumptions and guesswork that are folded into such top-down direction, the concentration of decision making into the hands of a few is most worrying to me. But even then, rarely do they decide exactly what piece of research each scientist must do next. In other countries, the staff in research agencies act as antennae for the science community, building consensus, and organizing committees that take the responsibility for decisions.

PEERING AT THE IDEAS

All countries have evolved systems to divide big slugs of government money into smaller pools for different areas, and further subdivide these into pots for disciplines. Researchers are asked to write proposals to access this money, justifying their importance and why it is worth funding. These research proposals are peer reviewed by several other scientists, normally from the same country because they know the local conditions and context of each applicant. Their comments are discussed by a panel considering a group of research proposals, and their ranking indicates which proposals should be funded. The funding agency uses this information to contract the institutions of the researchers to carry out the research. The aim is that ideas emerge from scientists across the research communities, and these compete with each other, with the most promising getting funded. The simple answer to what research gets funded is what sounds most compelling to the community of scientists in that country, compared to all the other available ideas at a particular moment.

Many variants have developed. Sometimes proposers get to see the comments from referees and can submit counterarguments. Sometimes the agencies insist research programs must be modified (normally made cheaper) before funding. A decision panel might visit teams proposing large bids in order to interrogate and confirm their strengths and weaknesses and explore their commitment. Proposals might be for one year or for ten years, for individuals or teams of a hundred. In some countries, researchers can hold only one grant at a time, while in others they may have many grants. Sometimes it matters how many of the promises made in your last grant were fulfilled; in other cases each proposal is considered on its own merits ostensibly ignoring past performance. Grants can be for highly specific research plans, or to fund researchers to do what they think will give the best new science, or for getting expensive new pieces of equipment or infrastructure.

Funding agencies like this system, since they harvest ideas from many researchers at one time, and the science community itself collectively makes decisions by delegating responsibility to rotating representatives. New ideas can be taken on their own worth, needing to convince only people who know the science areas and can make sensible judgments about quality and impact. Resources can move quite rapidly between different areas depending on their promise. If high-level discussions reveal a paucity of research in an important area, "calls" for proposals in this specific area can be issued. Rarely do these go unanswered by scientists, who always see how a call can be deformed toward their own ideas.

In this way a balance is found between continuity of funding, and the nimbleness in exploiting new discoveries. The model is that of market capitalism, with those scientists writing successful bids prospering and others left to blow in the wind. Nobody is told what to do, but the search for success in obtaining grants strongly influences what research is done. The view of the topical communities is sought doubly, in how well the applicants' previous work has been received, and in how interesting and feasible that they think the new work is.

WRITING FOR MONEY

Since competition for research funding is intense in every country, research proposals are honed endlessly. They have to appear novel and interesting but also practical to achieve and worthwhile even if some parts don't work. The referees are bound by confidentiality and have to acknowledge conflicts of interests, but they get to see my best ideas that I have not yet realized—I need to be careful what I disclose. In return referees have to write reasoned but anonymous critiques, useful to discuss with a wider panel of scientists who are not as expert in my particular subfield. Many scientists tell me they are justifiably paranoid about competitors stealing their ideas, but in truth that has not been my experience.

How to communicate ideas is again vital. Brilliant science wrapped in a humble or rambling account often generates little excitement for referee support. Too much vision and too little practical detail looks like a pipe dream unlikely to lead to results. Too much money requested, and the rest of the community (my referees) feels I am becoming greedy. All the talents needed to get high-impact papers accepted are also demanded for successful proposals. Many researchers spend more time on writing their proposals than on writing up the actual science that comes out of them. Each subfield has a consensus view about what a good grant looks like, and woe betide if I step out of this strait-jacket.

Part of the skill for gaining funding is encapsulating a piece of science into a program that appears to be achievable in a few years with a limited number of young researchers. Proposals are tales about how science might progress, evocative stories of different flavors that referees are asked to believe. Experienced scientists often progress novel research covertly, and then afterward ask for the money to do it. It is always easier to motivate excitement in something new, so asking for funds to continue working on a stubbornly thorny problem is not likely to shine. Competition exacerbates the push to claim attention, profundity, and novelty.

In my view this selection favors those capable of writing captivating fables with compelling narratives, but not necessarily the best research programs. However there is as yet no compelling alternative to this peer review of grant proposals for choosing what science will get done. Despite the profusion of problems and skews built into such a system, the alternatives do not have much going for them either. A finite funding resource has to be rationally rationed. Alternatives include giving successful older scientists large, stable funding over a long time, allowing them to do whatever they want. However, mutual support between elite scientists reinforces whatever structures emerge, making it hard to reduce weak-performing activities, and keeping funding productive.

Peer review is supposed to bring new ideas to the fore, reward excellence coming from any person, young or old, and deliver resources where they are needed. Detracting from this are strategic and practical concerns. Scientists working from this scheme cannot be sure what they will be doing in a few years. They hope it will all work out, but they remain constantly on tenterhooks to learn if their scientific appetite can be fed or if they face a period of starvation. To remain on this knife edge through a whole career span doesn't engender gentle mellowing or elder statesmanship. The science ecosystem thus becomes increasingly an experiment in raw selection pressures. Will the profound science experience of a mellow character prosper less than a Young Turk's drive to succeed? I suggest that emphasizing competitive traits will decrease diversity of scientists, and the capability of the whole system. Will mentoring and support disappear? Typically competition reduces the time people are prepared to devote to these. Which loyalties—to group, discipline, department, university, society, or country—will wither? Competition reinforces tight loyalty, reducing the efforts to wider support and encouragement across science. The science ecosystem is tested by all these pressures.

Other difficulties with funding by peer review relate to the increasing financial demands. More proposals submitted puts more strain on the same community reviewing them. The increased time spent in writing proposals is wasted (partially) if they do not get to be funded. Those trained scientists who cannot get funding are discarded by this system, embittered and often cast adrift from the research front line. A system like this is based on short-term research contracts for early researchers and translates directly into career problems (see the next chapter) since they similarly face uncertain futures based on remote funding decisions. Inevitably inequality increases in such a system, with successful scientists being more likely to get more grants (the same cumulative advantage as in publication impact), and with concentration of grants in elite institutions and groups, leaving others starving. Because this competition is so well defined as a clear race for money, it can become the raison d'être for scientists' existence, rather than just what is needed to develop resources to actually do science.

Trendy ideas in bandwagon science are more likely to get funded, even if they are weaker—in these areas political lobbying is more active. Since these ideas more likely directly compete with funded bandwagon research in other countries, they are less likely to advance the science web on their own. This highlights the challenge of deciding between completely different pieces of science, both of which are very strong in their own way. It is even harder when research straddles a number of different subfields. Such interdisciplinary proposals have a particularly hard time, landing with referees competent in only some parts, and facing panels who see money potentially "leaving" the fields they come from. The pecking order of science is brought to the fore, engendering a tribal state where subfields in a country band together to support each other's work, and mutually increasing their prospect of getting grants.

At a higher level, funding through peer review lacks overall strategy—the best proposals in one year might favor specific parts at the expense of the whole, blown by bandwagon winds. Another feedback arises when a subfield within a country favors a particular focus and approach, supporting funding in only their area, reinforcing people only working on that goal in that way, who in their own turn maintain the same focus. Shaking such an insular subfield out of its self-reinforcing loop is extremely difficult—outside intervention is threatening and unwelcome.

One accumulating view trickling out from the ecosystem of science is that the peer-review system is being stretched to breaking point. This is particularly directed at the grant-funding system because the increasing number of research grants submitted to governments, outstripping increases in budgets, means that the probability each grant application will fail is also increasing. By far the largest cost of submitting grants is the time and effort put in by the academics (which is funded by their universities, and often indirectly by governments). Current success rates of only 33 percent (UK, Germany) down to below 20 percent (the United States, other EU) represent a massive wastage—sometimes more than nine out of ten are rejected. Why do scientists bother then?

A key insight is that the cost of writing and submitting these grants is much less than their expected return. Although success rates are not high, the value of each successful grant makes up for this (in the UK, at least until rates drop below 10 percent). A fully competitive ecosystem can thus still get very much worse. A 2006 UK review found the net cost of academics' time used in reviewing and ranking each grant is only about a quarter of that spent in writing each one, presenting a system where creation is supported more than evaluation. These numbers worried the UK research councils enough to experiment with trying to reverse the rise in grant submissions—penalizing researchers whose unfunded grants are repeatedly ranked by panels toward the bottom. Naturally a storm of protest followed, however a quiet majority saw this as needed to avoid demoralizing researchers when success rates fall below 20 percent. Rather than deplete risky research ideas, it does seem to have encouraged fewer but larger grants (though I worry this is at the expense of the diversity of ideas). The danger of low success rates is that demoralized researchers notice better-funded parts of the world, and top researchers are lured abroad, disrupting national strategies.

WHICH DISCIPLINES SHOULD GET MOST?

Beyond these challenges, we should question what rationales exist for splitting up government funds between different disciplines. How much biology ought we to support, compared to how much particle physics? Despite scientists' claimed desire to avoid politics, all allocation of resources is inherently a political decision. It seems no one has the slightest idea how to handle the question. Particle physicists point not just to their explorer status, but to the spin-offs they generate on the way, claiming the invention of the World Wide Web at CERN. But spending large funds in any area with technological challenges is likely to conceive new things. A question to really ask is, given a large chunk of funds to put in any discipline, which one is more likely to generate promising outcomes for the taxpayer, and on what timescale? This would allow a rebalance with strategic views about minimal levels in each discipline. Instead for now, historic distribution patterns between subject areas are set in ice, mainly to avoid endless arguments with little recourse to any rational measures. Some correlation exists between the availability of research funds in a discipline and its number of researchers, but there is often no direct strategy for this. Such distributions become self-reinforcing and repel anything that might destabilize them, such as interdisciplinary research that blurs their boundaries. On a global scale, there is not the slightest attempt to define how many chemistry researchers the world might need, or whether we should limit the numbers working in the same subfield. We will return to such questions in the final chapter.

In many countries, different parts of government end up competing to fund the same science, for instance in programs for basic science, energy development, military, environment, or health, and these can operate both countrywide or in regional governments. Given that real breakthroughs in science are rare, these funders strongly compete with each other to fund the best prospects—giving power back to the scientists. No large bureaucratic organization wants to feel it is missing out on the next best thing, something that might be crucial to its mission, and their herd instinct is a powerful force. Proposals to each funder go through similar peer-review systems, and not surprisingly they end up funding the very same things.

In the science ecosystem, funders and disciplines compete for resources to feed the hopeful scientist. As in the stock market, confidence and feedback play a much stronger role than ideal, producing bizarre jerks and flows instead of steady growth. Sustenance lurches from one extreme to the other, famine and feast, in endless succession.

SUCCESS?

As taxpayers we are funding the successfully top-ranked proposals. We might be happy to go along with the idea that this is the best way to invest wisely in the most promising new science. But how do we know afterward whether it turned out well? We ourselves can't look at each piece of science, and even funding agencies find this an impossible challenge. These agencies are more focused on how to spend the next chunks of money wisely, not increasing the costs of their administration by evaluating the last ones. They have even less interest in highlighting what might turn out to be the poor decisions on their part, or of trying to second-guess what might have happened with different choices.

When scrutiny is closely focused on grant performance, the experiences are not often agreeable for anyone. The EU has a voluminous review process, bringing curses from the lips of a generation of scientists across this continent as they sink under bureaucracy. These reviews include evaluations midway through all grants, which are supposed to allow termination if there are major lapses—in practice, this is extremely rare, partly because scientists do at least the minimum needed to keep administrators at bay, and partly because the referees brought in to provide evaluations are from the same community of scientists. The review processes necessitate additional reports and presentations, many recycled from papers and conference talks, but recast in the formal schemes of these projects, and decorated with garlands of "milestones" and "deliverables." Such borrowing of formal project planning into science is supposed to better manage progress and resources, for both funders and scientists. But the uncertainty of making progress in science research makes such precise plans completely inappropriate in most scientists' view. They are mostly seen as fictions devised to satisfy administrators that research is progressing. Really they exemplify the low level of trust in the scientists themselves. While not completely unhelpful, these tools are applied formulaically and without any understanding of the science itself, as if these two layers could be decoupled from each other, and as if managing science was like any other spending.

The biggest issue is that research is supposed to delve into the unknown—without a significant risk element it should not be done at all. The vast multibillion-euro project at CERN to collide protons into each other at the highest possible speed might not find anything beyond the Higgs boson, and despite all the optimism of the scientists that this would still be important, they would join the public in being supremely disappointed. But without the lure of the unknown, we would not consider doing it at all. On the other hand, the experimental fusion reactor under construction in France is supposed to have as much risk pinned down as possible, and the funders expect a core set of results that will tell us how to make fusion work as a power source. Such scrutiny is not very different when a venture capital company is about to fund a start-up—funders urgently ask for answers to questions that can be answered only after the investment has played through. As individuals it seems we are happier with higher levels of risk than we can agree on collectively. We happily make the choice to travel on roads each day but demand governments eliminate the far smaller risks in airplane travel. Most of us taxpayers are comfortable with science being risky, because we have seen it succeed repeatedly in changing our knowledge or technology. But those we charge with managing the process find it hard to replicate this level of risk. They have to show that science succeeds, even though it is impossible for them to judge how investing in another research project might have given more.

The need to demonstrate responsibility is thus obscured by lower levels of detail, with funding agencies focused on the legal precision of exactly what funds have been spent on. Contracts come with flocks of provisions directing allowable ways of using money. This appears to have some sense, as we don't want to see scientists running Mercedes on their research budgets. On the other hand scientists are faced with highly competitive funding and the last thing they want is to waste their preciously fought-for resources. Different countries have diverse levels of accountability depending on which agency parts are on top, the financial accounting arm or the strategy science arm. This leads to situations where a scientist with money still on a research contract is not allowed to spend it on equipment instead of hiring an experienced researcher, or vice versa, even if it leads to better science. This wrestling over resources comes down to disagreement about who owns the resource: is it still the agency, the institution with the contract, or the scientist leading the research? Funding awards must be spent by a particular date, rather than used up at a rate dictated by the research itself. "Year-end," or "grant-end," are hypothetically crystallized moments at which resources are supposed to have expended according to some plan. That scientific research rarely follows any such plan generates frustration because it makes no sense to scientists, who have to use pots of money that otherwise would be more wisely spent slowly and efficiently. From the funders' point of view, these restrictions ensure that projects proceed at a suitable rate, and that the agencies can demonstrate activity to their ministry, sadly only easily measurable by the rate at which money is spent.

One reason that scientists seek diverse funding sources is to combine them to bridge the efforts of doing science. To deal with the restrictions on use of taxpayers' funds, scientists need other "unrestricted" funds, such as from industries who want a project done without caring about the details. In countries such as Japan, such unrestricted funds are judged dangerously maverick and ostensibly forbidden. Unfortunately each funder brings its own complications. Charities believe that university infrastructures are the responsibility of governments and make sure that they cover only the direct cost of research that they want done. But what is the true cost of research? This also turns out to be contested ground, as I discuss next.

UNIVERSAL BEASTS

It now seems natural that much science research is housed in universities (or institutes allied to universities). Permanent academics there have a role in teaching as well as research, including training young scientists. Without universities it would be much more difficult to learn how to become a scientist. Universities are also supposed to be insulated from the direct pressures of commercial imperatives, or direct research goals. This cloistered interior is supposed to be conducive to the deep thinking and long-term research required to develop entirely novel ideas (and attack sacred cows). In the modern world of science research, how do these aims hold up? Universities retain a stranglehold on bestowing the professional postgraduate qualification of a PhD. But the competitive nature of science means introverted mulling over long periods is extremely rare, though still crucially protected within universities. The idea that the perfect collaborator is to be found in the next building has also been eroded: the size of universities means that most scientists no longer bump into colleagues from a different discipline by chance, and most don't even know what all the academics in their own department are working on. But many universities have excellent infrastructure and reward successful scientists, so they remain a desirable home. The proximity of young students in each discipline breathes life blood into the contemplation and practice of science, in return for which they glimpse boundaries of knowledge, fuzzily distant from their textbooks and lectures.

Universities have taken on the role of housing science, and grown vastly in doing so. Similarly to organisms, they respond to pressures from inside and out. Their scientists always want more infrastructure than can be afforded, and to take as many young researchers as possible. Their departments want to satisfy the needs of teaching undergraduates, but then to shine internationally as luminous beacons kindling the future flames of their disciplines. To do this they need more space, more support staff, and to hire the best scientists in the world. Faculties want to prevent their departments from wreaking future financial havoc, and to attempt to balance mastodons and mice, the smallest and the largest. Central leadership wants to enhance the importance of their university and their place within it, and to drag the institution in necessary and promising directions. Given resources, universities will grow infinitely, although different countries vary their control, depending on external oversight and funding.

Many scientists moan about the central administration of their university, feeling irritated and fettered rather than enabled by their attentions, even though research would be impossible without this entire infrastructure. Ordering the bits and bobs for the next experiment requires an organization that attempts to get best value for money by negotiating centrally, but that procures it fast enough. A research contract might require a new lab to be kitted out, with gas and water piping, air-conditioning, and electronic environments suited to the program. Hiring young researchers requires conforming to many pieces of legislation in all countries, writing the formal contracts, and making sure that they are supported in starting off and in future careers advice. Joint research endeavors of large scale require all parties to cosign legal agreements that allow their commitments to be counted on by all involved, and negotiate for large purchases. To support the administration, buildings, accounting, and personnel components of a university requires funds termed "indirect costs," which are not tied to specific projects. Different countries fund these indirect costs in different ways, some in a yearly chunk while others as an extra percentage of research grants. Traditionally governments fund universities, but they find it difficult to control the demand and use of such infrastructure resources. Who decides which universities get more or less of such support?

The manna of indirect costs arriving with research grants (in the United States, UK, or EU) gives universities an interest in taking more research contracts, because they can build a larger pool of support staff. Each funding agency (and sometimes each university) has a different way of accounting for these indirect costs, which range from zero to more than doubling the direct costs of the research. Because universities in different countries are supported differently, this skews the competition for international or industrial funding—and it seems unlikely that a common scheme will be adopted. In countries with strong central governmental funding direct to the center of universities, there is less pressure and indirect costs not so sought after, allowing them to charge less for research costs. As with multinational companies, the larger the scale of the institution the more it builds state-of-the-art infrastructure, and the more it can attract the most successful scientists, so that its visibility expands. This relentless concentration is perhaps resisted only by science itself, which can emerge from the most surprising and ill-funded places. However, luring grant-festooned academics into one's university is seen as a key goal, and inequalities in universities cannot but rise further.

As universities expand, they find a situation where it is not easy to retrench when times get difficult. Universities are not very resilient if a country or state suddenly reduces science funding, or endowments from the stock market crash (for instance Harvard after 2008). They rely on the fact that no government wants to let a major university go under, and so they avoid insuring themselves against the worst cases. Rather like the overexpansion identified by Jared Diamond in *Collapse* as a root cause for societies destroying themselves at their peak, strong universities seem to have nowhere to go but bigger. Too big to fail is not restricted just to banks, and there is little oversight of these behemoths. The self-reinforcing effects of growing administrations leads to gradual distancing between scientists and support staff, so that neither side really understands how the other is trapped inside the science ecosystem. Managing this friction successfully is one of the challenges for universities.

In some countries the rapid expansion of science investment has created institutions that have yet to crystallize. Nascent universities and institutes such as in Saudi Arabia, China, Singapore, and

Figure 7.5: Old and new universities, founded in Cambridge (England, 1209, *left*) and KAUST (Kingdom of Saudi Arabia, 2009, *right*). Image credits: Bob Tubbs/Wikimedia; Sam Fentress/ Courtesy of HOK.

Russia have been developing hybrid cultures, unreeling global connections and attracting talent for a decade but have found it hard yet to leapfrog to prominence. In other countries, the university pecking order is ancient, acquired over centuries, and pits nimble entrepreneurial younger endeavors against slow-moving giants. In science, there is a role for both these types of hosts given the range of researcher personalities, types of science, and funding demands (figure 7.5).

So how do these influences change the science that is done? If a scientist wants massive infrastructure their grant is unlikely to be funded if they are at a small university with limited resources. Referees and funding panels discuss how much sense it makes to do a piece of science in a particular place. The scientist needs the backing of their administration, which will have to choose between different science paths if they are to create systematic expertise and reputation that can bring in further resources in the future. Accepting infrastructure brings a long-term burden to its upkeep, and universities are forced to balance this financial load against gains in prestige. It is hard for a smaller university to be outstanding in all areas of science, so they adopt strategies to gain distinction in a few select science-led enterprises, aligning their research institutes with these particular themes. Scientists who do not conform to these central initiatives are unlikely to prosper. The type and strength of young researchers a scientist can attract to work with them also depends on the university they are in. This makes it difficult to do some sorts of science in some sorts of institution.

As taxpayers, where would we like to place our trust in this system? We mandate funding agencies or charities to get the most for the least, while thinking also about the longer term. This is translated into a system efficient at distributing funding and demanding that it is spent rapidly. Universities accept the bulk of research contracts necessary for their survival, and to enhance their reputation. Scientists accept the culture they find themselves in, devising strategies to best get resources to do science. They also accept the culture of hiring short-term researchers, repeatedly applying for funding in small chunks, and are complicit in the evaluation of science ideas in the system of ranking proposals. Our trust is thus placed in a competitive framework based around universities, which seems to me increasingly to exacerbate outcomes that are not so helpful to our society by reducing diversity as described above. It is even perfectly possible that dystopian futures could encompass decision making through public reality shows that pick between glitzy science projects, or that university consortia bridging across countries could bid for major fractions of government funding for all research in a particular subfield (see chapter 10).

WHERE TO DO RESEARCH?

An alternative direction for many researchers is industry, which provides a research environment focused on delivering improved technologies while making a profit. The larger the company, the more opportunities exist for stepping further back from the most pressing technology concerns, to larger science questions that determine future success and prosperity of that company. At Hitachi, I was involved in constructor projects that had a twenty-five-year vision. So it is not adequate to claim that applied research is done in companies and only pure research in universities. A number of companies have impressive records in opening up completely new areas in science. But it is true that the ecosystem in which scientific research takes place is often differently balanced between industry and universities, with a number of other bodies such as institutes, trade bodies, and technology centers mixing variously these elements.

WHAT SCIENCE ADVANCES ARE MADE IN COMPANIES?

The last fifty years have shown many examples where basic science development has engendered completely new technologies. We already discussed the semiconductor camera chips developed by Kodak, but well before then, Bell Labs in New Jersey in the United States pioneered the electronic transistor for computing. Humans have now made more of these electronic switches than there are ants on the planet (in fact four thousand transistors for each of the 320 quadrillion ants alive today). More recently in the 1990s IBM invented the use of ultrathin metal layers for sensing the weak magnetic fields zipping by from each stored bit on a hard disk—this has driven the continued improvement in information storage density, although even this has a hard time keeping up with our demands for fast search and cloud computing. In many other areas too, such as gene sequencing, chemicals, biotechnologies, and materials, companies have developed some of the basic science that has led to both new technology and to new science. These advances may directly open fundamental challenges, or they may provide new tools that enable science to progress.

The answer to who chooses what science is done within industrial contexts is that it is a combination of top-down approval from senior managers, and bottom-up exploration by researchers working alongside their bread-and-butter technology projects. The areas of science tend to be much more grounded in materials, the stuff we use to make and build, to sense and probe. But because science risk is unmanageable, not only failures but wild successes can arise from any investment. The increasing awareness of risk has led to changes in the last decades with much less basic research funded in industrial labs. Instead a preference for watching ideas and technologies emerge in universities and small start-ups has developed, which are then pounced on by cash-rich industries. Driven by competition, in new technologies, in short-term shareholder value, and in management pay related to short-term performance, these have depressed investment in underpinning science, which has only longer-term payoffs.

SCIENCE WITHOUT STRATEGY

We have seen that what science gets done is a complex result of many interactions within the science ecosystem. What emerges is a mix of the trends within communities of scientists with the challenges identified by various more-or-less charismatic leaders, blended with feedback from public and funders, and randomized by institutional variety, national particulars, and individual whim. No one directs science. Overall strategy is remarkably lacking. Feedbacks similar to the publishing scio-sphere are ramped up further, with influences on funders coming not just from within science but from governments, industries, and the public as well. Competitions between funders, between industries, between governments, between institutions, their departments, and their faculties all combine to create an ecosystem that is constrained and increasingly aggressive.

In our growing science ecosystem, the challenge of creating major impact encourages ambitious scientists into bandwagons that help them project their messages further. This tends to increase funding focused onto such fashionable subfields. The resulting loss of diversity reduces resilience, and it is less clear that science productivity is improved. We will consider more conclusions for science funding in the final chapter, but what makes all this work better than it might seem are the people involved. The final element of the science ecosystem we thus need to turn to are the scientists who make it all work, who both turn the wheels and are trapped inside the cogs.

8

WHO BECOMES A SCIENTIST

In this chapter, I will look at how a scientist's career is built, and the selection pressures from the ecosystem. It will become clear that early stages are fraught with instability and fragile dependencies. At the same time, the memory and mentorship of the system depends on the older generations who have been relentlessly selected within it. First I will discuss how our youngsters become interested in science and different science careers they might encounter, including industry. Then I will focus more on the competition to succeed as an academic researcher, to gaining a permanent position. I will explore how universities decide to hire scientists and in which subfields, and how their careers and interest evolve subject to pressures from the science ecosystem.

WHO BECOMES A SCIENTIST?

In a straightforward world, if you grow up with an abiding fascination in things, in taking them apart, in asking the endless question "but . . . why?" and refuse to be satisfied with answers even as you grow older, then you are ripe for the community of scientists. Add to this a good dose of systematic education to inject tools of mathematical manipulation, organized facts and problem solving, and translation of the world into questions that can be scientifically addressed, and you can launch toward this goal. Filter your cohort through the sieves of ability, intuition, perseverance, and focus, and voila! But of course career progression is shaped by all the ecosystem actors that we have been discussing, and more. What you might end up doing in science is extremely unpredictable.

CONTINGENT HAPPENSTANCE

We treasure the idea of choice and free will. But without the duplicitous reinventions of hindsight, people's paths into research can look very random. Almost all scientists looking back later in their careers freely confess that events have buffeted them from one direction to another. Instead of careers speeding along the autobahn, their paths snaked through thickets and canyons, lurching around the rolling landscape of possibilities, and getting trapped in various pools of rewarding endeavor. Only those who stumble across a major result, one that ends up sustaining their research direction over a long time, appear to have a linear career. But these are often the people who we hear about from their recognition with awards. Because of our human urge to link ideas of progress with human stories about their discoverers, we edit out the reality of most careers built on "contingent happenstance."

This is probably a good thing since following hopes feels quite perilous. Younger scientists are even less experienced and more influenced by the cacophony of influences than the rest of the ecosystem. If young talent took on only the very trendiest areas, the diversity of science would shrink further, and with it the number of possible discoveries arising from serendipity. This is a major problem for the science ecosystem. With choice comes profligacy of resources, because many researchers flock around the same research ideas. They repeat research in more or less the same ways, not knowing how many others are simultaneously working on the same problem, or making slight variations that lead to little real progress. This wastage is part of the system.

An illusion of choice can haunt young researchers. "There *is* a right choice to be made about what I work on for the rest of my life. And if only I can spend more time thinking and finding out, then and only then will I make the perfect choice for my future." Delaying choices postpones the fearful decision to specialize, focusing on one option at the expense of all others. But the reality is that there is no perfect choice to be made, since so much depends on the way that science and its ecosystem unfold over the next forty years of a research career.

So why do we not instead take a pool of potential scientists and assign them randomly to research fields? This would ensure diversity, reduce wastage, and explore unfashionable areas. It would avoid the disappearance of talent from calmer fields when free choice leads to a mass exodus for new trendier areas—often the previous field suddenly reemerges in importance. High-frequency radio-wave science was seen by the 1990s as a backwater after the Cold War demands of ever better radar but has returned to dominate with mobile wireless technologies and optics of metals.

The idea to reduce the free choice of a scientist's research field in order to benefit the wider ecosystem seems a nonstarter. Who decides the current pool of areas from which to choose? How do we decide the boundaries between research areas, and police them? How do people feel inspired if their field is imposed on them? Instead these difficult choices are more subtly imposed using the influence of funding. If a scientist is fascinated in an area that is not trendy, then it is much less likely that they will find research money, or that their results will be published in high-impact journals. And since these are the indicators used in hiring researchers or not, it leads to reinforcement of the current alignment and direction of research. So to our list of attributes needed to become a scientist we have to add their fit to the science ecosystem, how robust their niche is.

As we have seen, the early science ecosystem used to be much smaller in all ways, and most importantly the feedback of the science echo chamber was weaker. This directly encouraged the growth of breadth and diversity of scientists that we inherited from the last century. A long-running perception of scientists is that they buck conventional attitudes to society as well as knowledge, even in their own science society. Now that we have greatly populated the science ecosystem and enhanced its competitive fabric, how well I fit into the ecosystem becomes much more crucial. To explore these influences I now take a look at how science careers are built.

CAREERING INTO SCIENCE

Typically, high school students interested in science go to university to deepen their knowledge. They are faced with an ever increasing choice of discipline options to focus on and after a few years have been exposed to a canon of courses that form the core of their discipline. Even in the face of increasingly interdisciplinary paradigms, the idea of disciplines is preserved. When you order up a chemist you know what you are getting—they master an agreed set of core modules, framing knowledge on which they accrete further studies. Structuring their mental space this way casts the perspective through which further information is viewed. The way a physicist learns biology is completely different from how a material scientist learns biology, or a biologist learns more biology. So there is no going back once you commit to a discipline, although there is always much more to learn.

By the end of undergraduate or master's courses, students will have experienced several months of research projects. This exposes them, often for the first time since childhood, to the notion that, not only do we not know answers, but no one does. For some this is a scary moment, but for others it is a revelation that there is no single "right" answer. Students who have aced academic factual courses can be dazed by the absence of reference materials, or appalled at the messiness of real-world research. In an experimental laboratory, results do not often emerge neatly. Instead, observations of partially understood experimental protocols, hazily followed, without knowledge of what are the key factors affecting results, drip-feed an intuition of what is going on. Simulation results might work only with artificially small volumes, have random errors that arise from an unknown source, and give "data" but no insight into what is happening. It is at such moments that students catch the contagious virus of science, or flee.

THE JOY OF SCIX

Science into the unknown (*SciX*) is addictive, in all senses. Those hooked early on science research are next set up to do a PhD, apprenticed to a professor. This is a "scary choice" moment since they will spend three to six years (depending on which science subculture they are in) mining deep inside a particular subdiscipline (figure 8.1). For

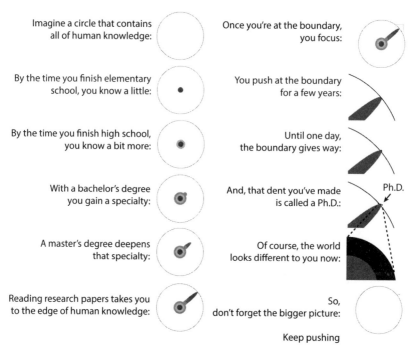

Figure 8.1: The terrifying experience of a PhD and the individual's contribution to world knowledge (reprinted by kind permission of Matt Might; see http://matt.might.net/articles/phd-school-in-pictures/).

each student, there are thousands of PhD projects that they could feasibly take up across the world—an impossible decision to make systematically. Asked by prospective students how to cope with this, I advise them to go with gut feelings—by who they will be working with and what the feel of their team is like. The whole point of a PhD is to learn how to do scientific research by the time-honored route of apprenticeship. One of the tough lessons everyone absorbs is that of a hard life and hard work with a thrilling payoff. This emotional ride means PhD students should have other reasons besides just science motivation to carry them through the slog and the low points—it is colleagues and friends who support us through miserable times. The converse is that at some points amazing things happen, your research really works, people you expound to are fascinated, and you feel the joy of creating something, the joy of the discoverer. People can get hooked on many different aspects of the science life—the building of equipment in the lab, the tweaking of advanced software on high-end computers, the community of researchers, the crafting of texts—but everyone feels the buzz of something working, the fizzing high of science intoxication.

The group choosing this PhD route is only a tiny fraction of our societies. Across the EU, the numbers of twenty-to-twenty-nine-year-olds who do a PhD varies from 0.7 percent in Sweden, to 0.3 percent in Italy and Spain, with the UK and France at 0.4 percent. These numbers correlate with the total number of scientists in each population, showing how the graduate student population flows onward to careers across research and development.

SUPPORTED LOCALLY, COMPETING GLOBALLY

The nature of the apprenticeship that is the PhD has slowly changed over time and generally is now more dependent on teams, infrastructure, and research funding. The picture of the lone researcher sailing under their own steam is incompatible with the sheer number of scientists and the volume and pace of research. Still, in most countries, PhD work itself is devoid of many direct influences from the science ecosystem. PhD students are insulated in a nest, most aware only of the need to produce outputs that have the highest impact possible in order to progress their science career. How much they should publish depends on their direction. While scientific papers are paramount in academic research, experience of innovation and technology are more important for starting industrial careers, with conferences and outreach more crucial for journalism or civil service.

For some, the highs of academia are not enough to sustain the less positive aspects of a research career. Postgraduates decide they crave something stabler, something that does not have to conjure results that are always new. It is different for lawyers or doctors, who support only their local communities, and are respected for doing the same things well. Scientists on the other hand are always having to do something for the first time and are always shown a global community who they have to impress: international, pan-disciplinary, and skeptical.

The training embedded in these PhD students, such as problem solving, model building, or deep thinking, makes them valuable in a vast range of occupations, from finance and management to teaching and government. More than 50 percent of PhD students move outside their specialized training to occupations that exploit their knowledge base less directly. In the UK, this compares with 80 percent of trainee plumbers who stay in the trade, while 95 percent of teachers stay in theirs. Whether this number seems high or not depends on what you see as the goal of this training. Many US academics believe a PhD is training for academia and so have a vested interest in prolonging their PhD students (who stay for five or six years) and treating outflows to other careers as a failure. On the other hand, the UK government rationale for funding PhDs is precisely that this outflow into all areas of society provides measurable returns (and not just financially). In Taiwan, bright students looking for rewards typically move straight into industry without a PhD, which is seen as a backwater. In whatever final career they end up in and in whatever country, asked how important their PhD training was to them, people stress the underlying skills they picked up and the important lessons they took with them.

INDUSTRY RESEARCH CAREERS

For PhD students who take an industrial position, their career path depends critically on whether they cling to immersion in science, or gravitate inexorably into management. Their decision is vital for many high-technology companies, who bring in smart researchers and then promote those with clear management potential as one of the most important recruitment channels for technologically qualified senior management. Apocryphally IBM funded its major research labs in the United States specifically to recruit handfuls of researchers into management, to become the future life blood of the company. Since they could take a view of the whole enterprise from science breakthroughs to technology, markets and services, they were better able to appreciate the full picture. This more than paid for the investment in research labs, with advances and know-how merely being the icing on the cake.

Those industrialists who continue to develop research play a boomerang role in the science ecosystem. As well as being measured against internal aims and the health of the technology base of the company, they have become important ambassadors. One measure of impact in science is how interested potential users are in your research. Industries are key potential exploiters of science advances, and the people best placed to understand how interested to be are the industrial researchers. So these people are wooed for all manner of reasons: to emphasize the real-world impact of research proposals, to enhance the prestige of a conference line-up of speakers, to articulate the importance of a research publication, or to signpost the relevance of an international collaboration. There are many positive aspects to this ambassadorial role, since such constructors really do get to see a wide range of research and can suggest promising avenues of application that may not just be their own. Their loyalties are often broader than those of academics, wider than a university, a discipline, or a country.

On the other hand, the limitations of money and time mean such industrialists have to choose which overtures to accept, and all the influence of the science circus is then amplified by their choices. Industrial researchers who can tell their line managers that they are spending effort on a collaboration within a recently trendy research area are more likely to be rewarded and respected. But often their real constraints are the imperative to fix a current technology issue in their company, not necessarily through understanding but through the quickest possible patch (perhaps not even leading to new science under the Frascati definitions). This can lead to frustration for young company researchers who have been earlier taught to think deeply without placing such value on expediency, but it provides a greater motivation for their forays into academic partnership. These tensions lead to cycles of collaboration, in which close links between industry and academia are alternately renewed and severed.

THE LONG LADDER TO ACADEMIA

FIRST STEPS UP THE ACADEMIC CAREER

For PhD students who are motivated to continue in academia, the competition sharpens. Half a century ago the next step on from student apprenticeship was to a permanent position, either in a university or research institute. However as the scope and ambition of each research program has increased, the demand for more experienced researchers has risen. As a result the postdoctoral research associate (PDRA or "postdoc") position was created many decades ago, typically providing two-year contract positions after the PhD. Nowadays most midtwenties to early-thirties researchers will have taken a few such jobs (sometimes more), often in different institutions in different countries, to broaden their experience and give them a strong enough track record to realistically apply for a permanent position. Scientists who want to be rooted in a specific place will find it hard to progress except in exceptional circumstances—gone are the days when a researcher never moved throughout their entire career.

It is all very different from most professions. Where else does one expect highly qualified professionals to work on short-term contracts that provide no guaranteed long-term career path? Medics slog for years of internship but know they will be employed. The science funnel pours the focused training of PhDs into the bottleneck of fiercely competitive permanent positions (figure 8.2). This yields an ultimately unsustainable career ladder in the science ecosystem. It is one that will survive for many more years though, since it offers the best prospects for bright researchers from developing countries to move to economies where their standard of living and life ambition is significantly improved. The model fits very well within the Anglo-Saxon capital-induced casualization of the workforce. It harks back to the itinerant travels of medieval stonemasons, who journeyed far and wide picking up new skills and seeing the world before settling. Ultimately the competition between people that is brought into sharp focus at the postdoc stage is decided by different societies' policies on immigration and the growth of education and wealth around the world.

The hierarchy is very blunt. When five postdocs work under a single professor's direction, a system that is not expanding will demand that all but a fifth of postdocs have to leave academic science. As with PhD students, how you see this exodus from academia depends on mindset: individual's stability, or society's success? In some fields that massively expanded in recent years (such as biomedical engineering)

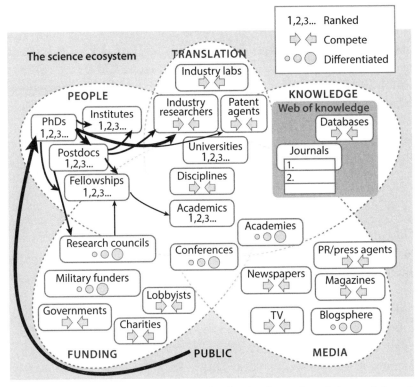

Figure 8.2: The scio-sphere of research scientists showing flows of people into and through science (by thickness of arrows).

the situation is far worse. A decade of trained researchers cannot get permanent jobs in institutions comparable to those they matured within—from 15 percent (the United States) to less than 5 percent (UK) of postdocs go on to permanent academic positions. On the other hand, very high quality research for society is produced by this cohort of experienced researchers. The destinations of these scientists are even wider and more varied than PhDs, spanning every possible industry, media, IT space, government, and beyond. Partly driven by their resentments of the system, or their realization that the rules are "everyone for themselves," they often start or drive small high-tech companies. Ejecting such scientists from their promised land encourages or even insists that they take risks, producing highly productive and innovative developments for societies.

There are dangers in a system that produces cadres of disaffected young scientists who leave the research arena without choice. Many do not look back fondly, though most retain a great love of science itself. Some become vociferous critics of nurturing and fostering research by government funding and the taxpayer. Finding a way to harness the great enthusiasms, insights, experience, and contribution of this cadre is rarely discussed. Rewarding the risk takers with self-belief to persevere does not mean we cannot smooth the path for those migrating to other careers, who can become positive ambassadors of understanding, particularly as lay interest in science increases.

THE RING OF FELLOWSHIP

One response to the oversupply of postdocs has been the invention of independent research fellowships offered by science funders. These split the fierce competition for permanent positions with an extra competitive level. Generally fellows do not have stable positions, but their salary is independently funded for a longer period (frequently five years or more). They are expected to produce high-quality (nowadays high-impact) research that grooms them for permanent research employment. This gives promising scientists early independence and lets them succeed or fail on their own merits. But they also gain power, as we shall see. Those failing to get fellowships recognize much earlier to prepare for forks in their career path and leave academic research with less injustice in mind.

For each such competition in the science ecosystem, we should ask how it works, and how its selectivity influences the progression of science. The winners of this competition form the germinating seeds that grow the next saplings of scientists. Biases in style or substance will permeate their way through the system in time.

What sort of person wins a fellowship? Postdocs prepare an application based on both their past research (performed at the direction of older scientists), and a detailed vision of what they want to do. Administrators group applications into themes and bring them to panels of eminent (to varying degrees) scientists who are asked to winnow a short list. Panelists are normally unpaid, and this work is not part of any job description from their employers, but although time consuming it is rewarding since it involves "real" people and mentoring the next generation, part of the long-term ethos of educational establishments. It also gives older scientists some insights into what research the next generation are interested in.

Panelists have great power to favor applicants from their own research areas—and adding scientists to any research field tends to increase its significance, resources, and claims on the media. Since fellows will mostly join existing research teams, panelists can sustain or hinder competitors or collaborators. More thorny influences come from science fashions—both applicants and panel are influenced by the media circus and "science bandwagons," highlighting research areas to focus on. In my experience, panelists try to be blind to the research topic and focus on the potential of a person, but negating influences is rarely simple. For instance, scientists in a trendy area will inevitably have acquired more impact and citations, although this may not reflect the quality of their science, their innovation in new ideas, or their development in science techniques.

Using experienced scientists as panelists is supposed to mitigate such influences, but it is not clear how. Such scientists have emerged from the science ecosystem as winners. Either they have ridden successfully expanding fields, or built careers as successful iconoclasts bucking the trends. In both cases they bring strong conceptions about the latest trends in science, for or against each one. Applicants themselves are even more submerged in the science hall of mirrors, so their proposals replay the mantras of each trendy area. Not surprisingly the influence of their current advisors is strong, inadvertently picked up in their team effort, or overtly in direct advice on their proposal.

Panelists must try to glean from submitted proposals how independent a young scientist really is, and how likely their growth into a major figure. Personal references for short-listed candidates are requested from scientists who know them well, and while these are important clues, they need similar interpretation for the reasons above. A reference from someone the fellow plans to work with is naturally going to be more positive than from someone leading a competing effort. Even where interviews are used, strong confirmation bias is at work in ranking the hierarchy of applicants.

Which candidates stand out through this process? Certainly the ones who are a bit different, with unusual backgrounds, or unusual research experiences—they may have been part of a planetary exploration mission, visited Antarctica for long periods, or studied colorful butterflies in the Amazon basin. How to compare their potential against a researcher who has worked on mixing colorless liquids hundreds at a time by robot and looked for the stickier cells that have grown fastest? After reading a hundred applications, panelists favor most those that stick in the memory. In interviews similarly, candidates who are articulate and who appear certain of their direction and future come across well. These are the confident glib ones. The deep-thinking but haltingly slow talkers, or the less polished, fare badly. Those who focus on their own research specifics, forgetting most of their panel resemble an intrigued public, come across as remote and too detail focused for driving any grand vision of new science. The media feedback has more effect here, giving scientists in trendy fields much more experience of delivering an appealing vision. Interviews may have a crucial role in gender balance as we will discuss in the final chapter. For myself however, I remain unconvinced this process produces an optimally diverse collection of scientists.

Although panelists take their roles very responsibly, they do not actually shoulder any personal risks in their decision. Funding is coming from governments, learned societies, institutes, or historic benefactions invested for the future. The panel's goal is to select people who are most likely to produce great contributions in future science, investing as wisely as possible. The funders for their part strongly want to believe that they have funded the best people— rarely possible to evaluate because their prophecies are self-fulfilling, with fellows inexorably pushed up the career ladder. So we end up with a cohort selected by as diverse a set of scientists as possible, and hope that the result is also diverse with the greatest potential for novel science breakthroughs. But note that the panelists do not have to live directly with the consequences—they will not be interacting strongly with their nascent protégés (a few in the vast pool). Most worryingly to me, if all the fellows selected have a research style that is aggressive, dominating, single-topic focused, and partisan, then this will be a style passed on to the future science base. Much research depending on collaboration would then disappear from the system, constricting research advances.

A JOB WITH AN AXE

A related system for academic careers is found in countries with a "tenure-track" system. In essence, postdocs win a potentially permanent position in a department, but with a major decision point after five years when they can be thrown back out if they haven't made the grade (accruing high enough impact publications and research funding). In Germany scientists have to become accepted professors by the end or move on. In the United States, while the number of tenured academics in physical sciences stagnated over the ten years to 2006, the number of nontenured academics grew by 18 percent. In the life sciences, the 11 percent growth in tenured academics over the same time was accompanied by a 25 percent increase in nontenured permanent researchers. In part this rise in the nontenured reflects the growth in the number of tenure-track positions who will not get a permanent job at the end of their grueling campaign to become noticed with accolades, attention, and funding. Such increased competition stresses the ecosystem enhancing pressures to survive among the many.

PERMANENT RESEARCHERS

At each stage upward, similar methods are used—competing for a permanent academic post is not so different from a fellowship. But one significant difference is that the group doing the recruiting come mostly from a single university department, and these people will have to live with the consequences of selecting a new long-term colleague. This time they take a real risk, since hiring and firing is extremely difficult in universities. Appointment panels are thus interested in a wider range of skills than previously tested: Might this person teach undergraduate students well? Will they be able to attract research funding? Will this person help run the department?

Still though, their paramount question will be, "is this person a good research scientist?" Recruiting an industrial researcher has a similar focus. How do panels evaluate this? Impact from the candidate's research is a strong factor, and references provide some way of evaluating the candidate's personal contributions to progress. However a single high-impact paper might be a lucky strike, a reflected fragment from a large team, or a long struggle in their own isolation. More reassuring is a number of strong papers across different areas, which starts to give confidence that this is someone who distils science from the messy edge of the research world, who selects promising problems of the right size to work on. This creates a strong bias to selecting people who work well within the existing system, who will move through the jungle of possibilities and reap enough novelty to publish well, and sustain their young scientists. Is this going to produce disruptive novelty with high impact that we, as funders, might want, even if only sporadically?

HOW DOES A UNIVERSITY DECIDE
TO HIRE A NEW ACADEMIC?

Any department allowed to recruit a new academic jumps at the chance. More colleagues mean less work for any one person, and generally more resources coming into that department. Larger departments hold more weight and power within a university, attracting more resources. Depending on the autonomy that they have, departments are thus another set of evolving beasts within the university ecosystem. Departments left to their own devices tend to grow with verdant fervor. The only restrictions are the resources of money and space. Successful departments attract more money and then garner more space, so success breeds success.

Rationales for hiring new academics seesaw between replacement and expansion. Most replacements occur as scientists retire, something that the ongoing changes in retirement age impact strongly, reducing the rate of new positions opening up (for the same staff funding). Even poorly productive scientists are rarely forced from permanent posts, so the opportunities for a department to move into new research fields by jettisoning one research group and starting another are cut down. Hence although partially autonomous, departments cannot move rapidly like companies but have a temporal turning circle often measured in decades.

EVOLUTION WITHIN ACADEMIC SCIENCE

HOW DO UNIVERSITIES CHOOSE AREAS OF SCIENCE?

Setting up a new group in a trendy research field is more likely to bring in money, prestige, recognition, and higher quality or larger quantities of students. In some countries, universities are independent beasts in the academic ecosystem and are free to hire in whatever research field they wish. In others they have to consult governments. Generally universities let departments decide which research areas to target, but against this central investment provided to a department for a new academic, they want reassurances that this is a smart move. The people overseeing such decisions within the center of a university are successful older scientists, and subject to the same media circus and science peer consensus. Different universities thus simultaneously try to build new groups in the same area. This leads to fierce competition for the highest-impact or most-visible academics in a new area—with a finite number of outstanding scientists, it is easier to get a permanent position if you are in the "right" field.

Universities are now more significantly influenced by their global branding, and how they are perceived. Driven partly by international league tables of universities, countries focus on becoming significant global players by creating large academic institutions that score highly. Science-led universities are thus becoming ever more responsive to prestige and funding, skewing more traditional roles (such as in Germany, where their training role in regional Länder is being subverted by research "Excellence Initiatives"). This concentration of resources into fewer large universities reduces diversity by suffocating other types of institutional cultures and styles. More positively it creates opportunities for resourcing unusual or obscure areas through the flexibility that comes from ever larger research budgets. The way such an ecosystem works out depends on who is making the decisions and what pressures they are responding to.

Is it efficient that individual departments are allowed to decide which research field to hire in, without reference to national need, long-term potential, or coordination? The current system often sees ten research groups in a trendy new area set up within a few years across a country, and these academics then all simultaneously apply for research funding. Early proposals satisfy the cravings of the science community and succeed well in funding competitions. By the time later proposals come along, even if they might be better thought out and stronger, the peer community is starting to weary of yet another bid in this area. The most restless scientists who move in first are not necessarily the strongest scientists. When science funding is so precious, this is a major problem—when excellent scientists do not find funding they are hamstrung in developing and rapidly fall behind a fast-moving bandwagon. So far, the Anglo-capitalist countries have generally felt that this sharp edge of competition outweighs any system inefficiency.

In other countries, there are national bodies to make decisions on how many researchers to hire in each field, passing on the undertaking to specific universities or institutes. In Germany, discussions within the Max Plank Society identify which area to set up a new institute in, and where to house it. In other countries such as Taiwan, the identification of urgent fields in which to prioritize academic hires is within government research agencies. These discussions among senior older scientists are again subject to many of the feedbacks I have discussed already. Whether a field in chemistry or biomaterials is more pressing depends on your view of long-term fundamentals compared to short-term prestige for your country or of the connections to current local industrial enterprise. The global health of science depends on a diversity of these decisions. However more and more such bodies look constantly over their shoulders at competitors and imitate the current trends. This leads to a situation where every country that I visit has an identical national list of priority research areas, with aspirations to be the best in the world. No committee risks bucking such broad trends or can ignore something competitors are crowing about. So conformity is strongly emphasized by these interactions, and all countries worldwide are converging in their research visions.

HOW DO SCIENTISTS CHOOSE THEIR SCIENCE?

Once gaining a permanent post, how do scientists' interests evolve? They will have indicated at interview what they plan to work on, to their new department, institute head, or industrial management. But in fact they will mostly have great freedom to evolve. Explorers will be constrained by the impermeability of research walls to answers they seek, so that a mirage of easier questions can seem attractive. For instance, many have run aground trying to progress the understanding of consciousness, so where to start anew? Sometimes these walls are inherent to the research, with insufficiently advanced tools or data. At other times these walls will be funding constraints, as in the evolving funding landscape of large particle physics experiments. Constructors face similar constraints. When the toolbox they have selected stops throwing up intriguing novelty but substitutes it with the graft of harder-won evermore complex details, they have to choose whether to combine in new elements or prioritize new visions.

In most cases, what bounds their envelope of possibilities is what can be actually realized with the resources available. Mostly this means funding for equipment, running costs, and researchers. Motivations for a change may come from the increasing difficulty of being further funded for what one has been good at. The funding ecosystem has a constant thirst for novelty, and it *demands* to be quenched. A researcher may develop several alternative research themes, connected enough to their previous work that they can argue risks are low. Whichever one is funded will end up being their future. This Darwinian evolution is subject to a landscape of research fitness that moves them more or less slowly across the plains of possibility. Some scientists circle for their entire career around the same places, making especially firm ground for a discipline. They populate and sustain an ecological niche that is just the right size not to die out or attract predators. For others who are easily bored, the lure of the unknown takes them always into the misty distance. They are hoping to stumble on buried treasure, unexpectedly located in science terrain not carefully picked over. As our tools improve, how we explore the landscape changes too, and previously inaccessible cliffs can be circumnavigated and rivers forded upstream to open up virgin science lands.

THE POLITICS OF GRAND VISION SCIENCE

Some scientists are strongly vision led. For the most part these are simplifiers, who can articulate a goal that remains static over many years. Here, a viable strategy is to establish fixed bridgeheads in the research landscape over long periods that accumulate resources, carve out territories, and suck in the large funding needed to grow ever larger. These scientists have to become fervent lobbyists, drawing all to look favorably on their cause. Since big science costs are equivalent to vastly many smaller projects, such teams have strong motivation to evade established funding routes. These teams have to devote resources to highly effective public outreach, winning influence as in politics.

Decisions of this nature are clearly political, and a major problem for all science ecosystems is how to balance big but expensive goals against many smaller but highly diverse research targets. The most common solution across cultures for such high-cost bridgeheads like cosmology, particle physics, cancer, or neuroscience has been to make such decisions explicitly political, separating out the funding streams for these areas and letting each separately bid for a share of the tax-base resource. Scientists on either side of the resulting divides grudgingly accept the other's rights, but at the same time (with an eye on influencing the politics) they make unsustainable claims on the priority of attention needed for their own areas. Often the argument is that because some other major country is now focusing large resource into a research endeavor, we have to do it too otherwise we fall behind. While personally true for the researchers involved, at a national level this is not an effective argument. If competing countries choose one area, is it not better to choose a different one? What is the rush to do this research now, why not in ten years? Not expanding expertise in one area retains and develops skills elsewhere. One challenge in choosing what to resource is the vastly different cost of research questions that appear on the surface equally simple to articulate. And as yet there is no way to really estimate the potential benefits to societies of choosing one grand challenge question to tackle over another, despite choruses of claims to the contrary. It is true that spending money on any project has potential to generate spin-offs. But the real question is what outcomes might emerge if the same large-scale funds were redirected to other science. I believe it is necessary for us all to understand the full science ecosystem, if we want to help our societies decide how to continue supporting science endeavors.

BANDWAGON SCIENCE

Well-funded constructors on the other hand, who generally require a much smaller scale of means, can jump into new fields at will by redeploying existing resources. In some ways, these scientists have the most freedom and best resemble the elite Victorian gentlemen scientists who funded their own endeavors. Such scientists can follow their nose in any direction. Rarely do the strings attached to research funding constrain the exact results that must be obtained (the whole point of research is that one does not know the outcome). As long as the impact of the research outputs is high, no scientist passing judgment is likely to complain loudly. This capability to jump onto a distant research peak is aided by the rapidity of technology advances in recent decades. Research groups used to occupy and protect their research peaks from invaders, their fortifications based on the complex home-built equipment they had devised or the myriad little tricks they had painfully accumulated over years of experience. This is no longer such a straightforward strategy. Most equipment rapidly becomes available to buy, since there are always young researchers cast out from the academic fortresses, who see an opportunity to turn their construction mastery into an alternative company career. Or you can buy know-how and experience by luring away a younger researcher, offering them the next step on the career ladder. Such sharing of know-how is now institutionalized, with many larger science institutions, such as in the EU, encouraging mobility between groups.

The lush virgin research slopes are thus open to migrations of constructors. However, they must be successful in rapidly producing novel results, so competition is especially high. Scientist parachuting in from faraway fields bring with them different approaches, alternative tools, or unusual pipelines to previous research that illuminate a new area in a different light. All this creates a rapid explosion in variety of research and a clamor for novelty within the trendy area. A "bandwagon" is created, but perhaps the word gives the wrong impression of a million souls plodding together through the desert landscape. Rather there is a feeding frenzy in dense rainforest, all hidden by huge trees. From outside lots of shaking foliage is seen, and a cacophony with little clarity. From inside, very little can be clearly seen.

Moving far from home is a scary business, and jumping research fields is quite risky. Who knows if the community you vault into will not see you as an outsider? Or withhold the nurture and encouragement that all individuals need? Will all the good finds have been made already, and leave you only picking over dead bones? Will the union you propose grow or wither—only rare combinations of science ideas give more than expected. Leaving a well-ordered community and entering a free-for-all can be daunting—I have done it half a dozen times with different experiences each time. In bandwagons, science quality becomes extremely variable, and there is less listening and more trumpeting. Many pieces of science are done at the same time in similar ways, and urgency trumps accuracy. On the other hand, truths are much more rapidly uncovered. Science is normally rarely repeated to check its veracity (because it costs valuable funds, and being second gains no glory or impact), but in a bandwagon it naturally happens, and the web of scientific results rapidly fills out. A young researcher in such a position has to worry constantly who is ahead of them. Research leaders have no choice but to exhort their teams to ever greater efforts. Good for science, but uncomfortable for the lives of many individual scientists.

Bandwagons are unstoppable in the current science ecosystem, given its stress on competition and its in-built flexibility. They arise from a number of conditions, including serendipitous new results that galvanize a community, and starvation of results in previously rich science seams. Sometimes a faction of constructors from a mature subfield jump en masse, upping sticks and bringing their entire community with them. This punctuated evolution of the science ecosystem is becoming more accentuated as the rise in number of scientists collectively work through new territory faster, and are more dependent on novelty to fuel impact and funding.

EVOLUTION OF THE SCIENTISTS

For me, one striking aspect of my early science career was the eminent established scientists who I was privileged to share time with, who exuded a passionate and apparently innate love of science, all displayed with little of their own ego. Often these scientists had cultural origins from Central and Eastern Europe, and their love of learning and teaching was what sustained them. They delighted in new ideas and were happy developing at their own pace, making time to encourage scientists around them. By contrast, the new promising academic of today will be a star in a trend-setting research field, supported by large research funds they have won for a number of different science projects that go in different directions, and thus with a large team that they manage. They will be thus in strong demand for international conferences, where they spend considerable time, and their contact time will be devoted to recruiting new students and postdocs, to supporting their bids for fellowships and future jobs, to pushing their researchers' projects forward rapidly, and to turning the results into research papers in as high an impact journal as possible. There is much less time here for thoughtful gestation, for mentoring of younger colleagues in the same department or field, for exuding calm and confidence.

We talk about the *health* of an ecosystem, or the *strength* of an economy. The latter has traditionally been condensed into GDP measures, but it is interesting to note the rise of "happiness" indexes for different societies that give more credence to ecosystem services, as in natural environments. Few attempts to measure the health of science through the happiness quotient of scientists exist—though one recent report shows scientist's satisfaction generally scaling with the overall happiness index of each country, led by Denmark and Switzerland (only French and Indian scientists seem distinctly happier than their compatriots). Happiness is strongly correlated with absolute per capita income, so perhaps we might expect scientists' happiness to relate to research income per scientist. As we have seen, this varies strongly between countries (figure 7.3) and has both risen (the United States, Korea, China) and fallen (Canada, UK). These changes indeed seem to account for some of the anecdotal experiences of scientists' happiness in different countries over the last fifty years. But we simply don't know if good science comes from happier scientists.

WHAT DO SCIENTISTS ACTUALLY DO?

Connected to happiness is the experience of what scientists actually do every day. From my viewpoint, the lucky ones are younger, at PhD and postdoc level, immersed almost continuously in the doing, the building, the puzzling, the wrestling with ideas. But on their minds weigh the insecurities of the future. More senior scientists like (more lately) myself write project proposals, write e-mails, write letters of recommendation, write and edit scientific papers or reports, write to argue with editors and referees and managers, present their ideas across the world, balance their finances, and battle with administrators, as well as joyfully discuss, argue, and occasionally do science with their research team. While the latter is the real science and the part that the academic was originally selected for, the ecosystem demands an intense focus on the former list, on the competition to survive in an ecosystem built out of many parts that interlock in a way to create a whole that is no longer utterly desirable. The key to being a successful scientist is keeping close to the active doing, the active ideas, and the active arguments. It is surprisingly difficult, given so many ecosystem pressures that rate success in other ways.

THE SCIENTIST EMERGES

The science ecosystem underpins this evolution of the scientist, and we can see how contested the space is to alter anything. Intense competition enables cavalier treatment of younger researchers. It also favors those scientists who are entrepreneurial, charismatic, articulate, focused, and passionate, just as in any career path. However this perhaps does not chime with a view that science progress should depend only on science ideas, which is most definitely not the case. I have identified two politicized areas particularly prone to maladaptions. Big simplifier science often bypasses cost-benefit comparisons with other parts of science. Conversely, bandwagon constructor science traps resources inefficiently.

To finish this discussion of science, I find it useful to consider an ecosystem analogy. Just as sunlight corresponds to the funding needed to develop science, people are more like the rain that fertilizes everything, the water cycle of the science ecosystem (figure 2.8). Seeded from the clouds of our societies, their science interest condenses as droplets, falling into the many watersheds of different science disciplines. Flowing down the headwaters develops their knowledge through the high slopes of the science canon. Cisterns and millponds collect people's expertise and science insight, siphoning them off to power the turbines that drive our industry and the agriculture that feeds our society. All this requires the dollars of sunlight, the energy that powers the cycle. At the end, we refine this ocean of knowledge, distilling fractions of science that infuse society with interest and motivation, and restarting the whole cycle of scientific rain.

At this point, I have completed our survey of the science ecosystem, but what is missing is a way to see all the tensions acting at the same time, and to feel its rigidity as a whole. In the final chapters I will tackle this challenge and ask what it might prompt about lessons for the future.

9

THE FUTURE OF SCIENCE

In this book, I have tried to map how interactions between different parts of the science ecosystem produce the tightly enmeshed network we have today. Although many aspects vary with discipline, country, and society, within the world of science these are converging. I think so far what I have described is recognizable to all scientists, but in these final chapters I will more personally try and broach the harder issues about what we can expect for the future and what we want.

IS SCIENCE IN GOOD SHAPE?

Most scientists are optimistic about science. Young researchers flock ever more to its siren call of intrigue and utility. As members of the public we are highly enthusiastic about science, expectant, willing to be enthralled and to invest efforts. Across the globe, the same fraction of children maturing into adulthood become fascinated in questions about the world around them and follow their curiosity. More funds than ever before are being spent around the world on science. But scientists embedded in it fear the racking up of pressures intrinsic to the current ecosystem. An increasing number seem dissatisfied about how their livelihoods work, they want to change things, but they feel they have little control.

I also want to focus on diversity in the science ecosystem, diversity in the ways of doing science. Many societies, types of institutions and companies, and disciplines support research. However the pressures of competition I have tracked across the entire ecosystem seem to be progressively leading to lower diversity: connectivity is not cost-free. For instance, perceived improvements in the way research funding is allocated between scientists are now rapidly copied and spread through the entire ecosystem. Similarly, emerging bandwagons sequester ever increasing resources since it is at these sites that competition focuses. While reducing barriers improves sharing of practice and knowledge, substantive change can become slower since radical new models are not explored (which often requires separately protected habitats). Are there ways to encourage different impacts in science without reintroducing barriers that waste hard-won resources? Perhaps globalization and growth then have hidden costs for science.

Ecosystem health attempts to measure the robustness and recovery capacity for an ecosystem. Disruptions can lead to collapse or to a rebound in ecosystem health, depending on the resiliency of its original state and the toxicity of the disruption. Is the science ecosystem robust with a strong recovery capacity? As we have seen, the system of science contains a plethora of compartmentalized competing units, all of very different sizes, complexities, and motivations. Their mutual competition plays out on different levels, with the dynamic balance at one level setting the habitat within which struggles on different levels are fought (figure 9.1). At the moment, I see these competitions only expanding. The density of most types of actors involved is increasing, while the diversity on every level is decreasing.

TENSEGRITY

The scio-sphere diagram shows a reassuring flow and fluid connectivity, but a more accurate view is of blocks held together with tightly tensioned cables (a sort of "tensegrity," figure 9.2). Each layer of competition (for instance between different universities, or

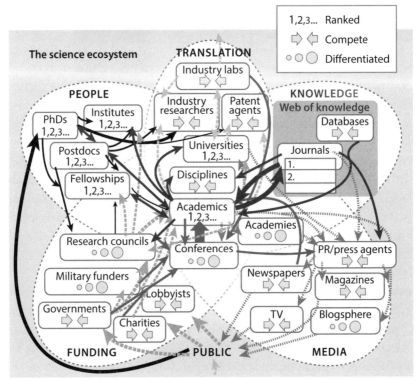

Figure 9.1: Scio-spheres, and the main flows of influence and tension.

different funders) forms its own incompressible block within the whole ecosystem. Just as the pressurized hull of an inflatable dinghy (or the inflated living cells in a plant) forms a rigid component, the intense pressures between each competitor in this layer produces a subsystem that overall hardly deforms. Each block contains a single tier of competing peers, and the ecosystem is completed by the tensioned interactions between them. The tier of academics is tied to the tier of funders, which is tied to the tier of universities, and so on.

Rigidity is built in by the tight web of interactions. In tensegrity structures, the floppiness of the whole disappears when the cables are tightened—just like bicycle wheels that have a solid rim connected to a central hub by spokes that pull in tension against each other. This is the image I am trying to convey of the scientific ecosystem—rigidity is produced on two levels: the competition within each component

Figure 9.2: Tensegrity structures combine rigid struts with tensioned cables (for *tensional integrity*), from Skylon (*left*) to modern bridges (*right*). The Skylon tower at the 1951 Festival of Britain, as designed by Felix Samuely.

that gives incompressible elements, and the tensions that link these components together. As the ecosystem has flexed around in time, it has gradually settled into local taut configurations that are now lashed by the competing tensions. This emergent rigidity leaves participants lost as to what to cut or lengthen to reintroduce the flexibility that is essential for resilience and diversity. If they do what is best for themselves, using marginal extra time or resources to achieve more, this actually increases both the rigidity and tensions. When individual strivings make a group worse off, we are in a situation known as "the tragedy of the commons," where satisfying everyone's own needs means for instance that all trees in a forest get cut down. Global action is then the only helpful path. But with science, no global path is obvious, and the system is becoming increasingly frozen into the current state.

In science, the knotted feedbacks are so taut that change will not be spontaneous. Evolution in the science ecosystem is a process of co-adaption with only small steps taken by each participant. The strong competition at each level provides very defined equilibrium states and removes incentives to shift even a small distance in a new direction. Each new science generation is born directly into the same web, and there is no opportunity for growing diversity. Despite nascent attempts, it is not clear that emerging information technologies are opening dramatically different ways for the science ecosystem to operate, rather than just increasing these tensions.

WHO EATS WHOM?

At the heart of stable natural ecosystems are pyramids of paired predator-prey relationships. Antelope herds are winnowed of the young and infirm by prowling jackals. Schools of fish are driven to frenzy by marauding sharks. Ant-lions patiently await the unwary ant. Where are such relationships in the science ecosystem, or is such single-handed combat a false metaphor?

For every winner of a grant, a major research paper, star publicity, or a new high-tech product there are many losers: research groups in competition that got lost along the way and failed to be first to publication, wanna-be-academics that didn't get a permanent job, papers that died in obscurity, or inventive ideas that never made it to market. While such competitors often go head to head in combat, the loser is rarely eaten by the victor. Instead, as in a role-playing game, the characters lose points and their fitness to survive declines. Wars of attrition are closer to those of stable societies of organisms than mortal combat.

Individual battles form the heart of the science competitions. Are there particular advantages for "hawks" or "doves," or for sharks within the scientific oceans? Can we say which ethical principles for action are of benefit for science, and if there are several? Consider the scientist who is highly aggressive and aggrandizing and wins a high-impact paper (the hawk), rather than delivering the same truth wound round measured words in a lower-rank journal by a competitor group (the dove). Does the stability of science rely on a certain ratio of these hawks and doves? In simple game theory models, too many hawks start fighting unproductively and there are not enough doves to help them out. Too many doves and a single hawk can go on a killing spree, rapidly reproducing. The stable balance preserves a small ratio of hawks appropriate to the details of the payoffs and interactions. If all scientists overplay their research, the system unproductively fails to guide us to what is really important. Too few science doves, and the research rate declines since they provide too few incrementally useful advances for mining by hawks to give breakthrough progress. This simple game scenario already contains many kernels of behavior that I see in science.

The same game is mirrored for leadership in an industrial lab or university department. Too many hawks, out for themselves only, makes department unworkable (as many a frustrated head has found). Too many doves, and the single hawk prospers unhindered causing havoc in their wake. Many of the rules and social conventions in our societies aim to hamper such untrammeled exploitation, freezing out those who don't play by the rules. But the personal rewards of global esteem in science can trump such local village etiquette, leading to hosts of dysfunctional teams. The higher the rewards, the more it can pay to dump your colleagues.

Another equivalent is the science journalist who sensationalizes their reporting and grabs the front page while their competitors gently emanate measured prose to the few delvers who make it to page fifty-four. A bigger science impact is delivered to us from the hawk, but at the cost of an accumulated swarm of confusions where science doesn't add up to a big picture but to a tribal conflict.

All these examples show that accentuating the competitive interactions within the science ecosystem tilts the stable balance. The new equilibrium may not balance where we want it, wasting resources, and turning up the clamor level of claims, without giving us anything back.

WHO SHOULD WE ASK ABOUT SUCCESS?

I have argued here a personal view about the science ecosystem, but, without more statistical evidence, who else could you listen to? It is not clear whom to trust. Our social human nature enhances our desire to hear personal stories, to know not just the science but also the scientists behind each discovery. Vignettes and legends recounted in the adventures toward a new piece of knowledge help lodge its science more securely in the mind. They also show science to be less serial and linear, more like the reality of happenstance and fateful caprice. To make the transmission of science effective, I showed in many chapters that science teams have to construct fables about how its secrets are uncovered. Successful scientists must be then also good fabulists and visionaries. How do we select who to talk to for an unbiased view?

Scientists who have amassed accolades within their field must clearly be good at research—rarely does blind stumbling in the dark uncover treasure. Star scientists have a natural inclination to respect the specific community that has lauded them—it reinforces that their work was indeed important. Since criticism, even well meant, is the order of the day in science, all scientists respond to praise with great respect for their peers—their past glories are now being authorized as important. It is through this prism of past importance that they can easily form a conclusion that not as much science being done today is important.

It takes time for the real significance of results to become clear. Unless I happen to start a bandwagon, which by itself wrenches the directions of a large number of scientists, mostly I cannot be sure of what a new piece of science might mean for at least a decade. Most of the scientists we recognize in society are now later in their careers—witness the age profile of any prestigious academy like the UK Royal Society or the US National Academy of Science. The scientists we consult are senior, past the midspan of their careers, and looking back. If we try instead to talk to younger scientists, our problem would be how to collect a representative sample since their opinions are quite diverse. Younger people are typically more optimistic, although more aware of the strains within the heart of the system. It would be valuable to do this legwork, but we would have a hard time giving credence to one opinion over another, or uncovering clear trends about science. Generally it is not done (but please join efforts at www.sciencemonster.com).

Senior scientists thus naturally look back fondly to an age where they see more of significance was done, their own work included. They ask "where are the new breakthroughs that overturn established theories?," "where are the startling results beyond current understanding?"

THE PACE OF BREAKTHROUGHS

We have seen that the amount of science underway now is many-fold greater than in the middle of the last century. When we question why there are not more breakthroughs that overturn large amounts of previously accepted knowledge, we are implicitly assuming something about the *rate* at which we expect such breakthroughs. We might expect that the more science we fund, the more breakthroughs we should get. However the evidence for this is not so good.

Each bandwagon has its own story. One for instance is the discovery of high-temperature superconductors, materials in which electrical resistance plummets to nothing below a critical temperature near that of liquid nitrogen at –200°C. This sparked off a massive research effort in the late 1980s, which was spurred by the goal of increasing the critical temperature still higher, up to room temperature. Not just new technology like levitating trains would be enabled by this, but also fascinating scientific investigations. Since the superconducting effects arise directly from quantum mechanics, here was a system in which quantum phenomena might be accessible to all. What is startling is that, even thirty years later, we still don't know exactly how these materials work. After so much effort, in fact we have many, many different theories about how they might be working. But the materials are complicated, a classic example of *constructor* science generating a new system that uncovers realms of new physics. (A typical conundrum for constructor science asks if we should say this effect was "discovered" or "created," since the materials are not found in nature.) These investments have encouraged a large clan of scientists to work on similarly peculiar and complicated materials, combining intricate, ordered layers and magnetism together with electrical phenomena. Funding the bandwagon did not give clear revelatory results but continuously inches open the door to new ideas, slowly leading to realizations about profitable directions to pursue. These new concepts have emerged not from the initial mass entry by a horde of wide-eyed scientists, but from those in for the long, hard struggle. In my view, almost all bandwagons share this characteristic of early-on wasted resources in the desperation for speed.

It is a peculiar question to ask of science, why it should continually rebuild itself, knocking out whole chunks of its structure in the process. When we find new directions emerging in the arts, such as the evolution of video installations or graffiti, it does not destroy what went before. Instead the old and new are used together in dialogue and in contrast to set in context what interesting questions and resonances can emerge. Somehow in science we have held onto the *paradigm shift* model attributed to Thomas Kuhn, that destruction is needed for progress and prior interpretations are inferior (certainly new artistic movements would like to wield this power). It is seen that the most fundamental science must emerge from knocking over what went before. Unfortunately the need for scientists to demonstrate novelty and impact means they are now pressured to present their work as destroying past science and rebuilding new paradigms—while in reality it is usually nothing of the sort.

A second feature is that the increasing churn of apparent novelty in innovative artifacts purchased by the public sets an expectation that science should progress ever faster. The argument is not clear as to whether fundamental technological innovation is speeding up—one thesis says that since evolution dominates processes that underpin progress, better and better routes will always emerge. We are led to think so, because it attractively increases our desire to purchase the latest gadget. But most lists of the highest-impacting technologies place their significant discoveries at more than a hundred years ago. Similarly science seems to have a pace all its own, which is not driven by the clink of coins. It is more like a pendulum, which has a beat that doesn't depend on how much you increase its mass, because how gravity pulls it in is balanced by how hard it is to turn around.

Why can we not accelerate or decelerate scientific novelty? The United States prides itself on creating a culture that is invention and science minded. By many standards it has succeeded, with still much of global scientific impact deriving from there. Why should some societies be more productively innovative than others, or is this just an apparent difference only? The same countries appear at the top of different rankings for innovation (consistently the United States and UK). Explanations range from cultural and historical, to the presence of elite institutions, an atmosphere of competition, the English language, encouragement of collaborations, or researcher mobility.

One reason that the rate of science progress is hard to influence is its embedding within culture and the science ecosystem. Many of the most significant discoveries depend on an alchemy combining new technologies and scientific advances from different areas, importing them into something that may have been studied but then ground to a halt in the past. Such advances require time. For the miraculous advances in one discipline to help another often requires at least ten years developing new technologies or ideas. The cycle of research requires communities to develop, publish, organize conferences, and swap researchers advancing through their careers, with fellowships and permanent appointments. The natural cycle of researcher growth is thus another factor that sets a timescale on science progress. While recent studies find nothing can systematically predict the performance of different countries, the influence of social networks is apparent.

RULING THE RULER?

As we invest more of the fruits of our economies in science, measuring what we get for it has become more important. The energy source, the enabling sun of the science ecosystem, is money. But we are increasingly using financial value as a way to directly measure the health of the science ecosystem as well, in a way that we would hesitate to do when viewing the natural landscape (even if it too is costed these days). The comparison of science to the economic competitive market, and also its increasingly entwined direct interactions with industrial innovation, encourages the use of money metrics and translation of impact into currency, with implicit conversion rates depending on the stance of the measurer. This helps in organizing and justifying allocations of resources, but it does not support science ecosystem services, nor maximize the public good of science. It has become increasingly distorting to measure all science from the simplistic monetary yardstick. We need a new diversity of measurements, to keep science itself diverse.

PEOPLE SUCCESS

Our measure of science has been in terms of new ideas and results, scientific output that has impact. But this misses the influence of people in science, and good scientists with a particular worldview can make more impact than the sum total of their published science. Richard Feynman was of his time, conveying charismatically the essence of a scientific life. We ought to be alive to the evaluation of scientific careers, sorting which provide a good return for society. Are scientists today what we want them to be? Undoubtedly it is not an easy life—competing against everyone for tantalizing rewards in esteem and partial financial rewards. It is a life of self-doubt and self-improvement, but it is a self-considered life. The towering mountain of hierarchical achievements leaves many protoscientists seeking alternative motivations but still delivers constant new blood to the system with no sign of senescence. My nagging worry is whether science careers still come from diverse enough nurseries—do they still emerge from patent agents (as Einstein), phone companies (Tesla, Edison), the shop floor or engineering yard, or the back bedroom (Alexander Bell)? The answer seems to depend: IT advances flower everywhere, but nowadays biologists are raised in conveyor-belt labs. As noted in previous chapters, if scientists are trained not only better but evermore similarly, our diversity in education reduces.

IT IS NOT LIKE IT USED TO BE

Identifying breakthrough scientists is getting harder—geniuses are sought but rarely agreed on. There are simply too many competing people working in too similar areas to allow the development of fertile new ideas in isolation over an extended range of time. Some scientists look back and bemoan this change. Where is my chance for greatness? On the other hand they never personally knew the earlier isolation, the struggle for resources, and the lack of career opportunities. Competition was still there in the past, because even a single competitor sees the potential in a great idea.

We are more part of a "hive mind" now. Often, advances are simultaneously uncovered by different groups separated in different parts of the world or ecosystem, similar to "convergent evolution" in which useful traits evolve in completely different species through different routes that hit on similar optimal solutions. But unlike in nature, communication is now global, and even if knowledge diffuses rather than ricocheting from one scientist to another, similar ideas are of their time. In other words, the culture of science opinion, what is important or topical and what is a sensible sort of science to do, is generally agreed. But this agreement is rather like fashion, depending on the number of hooves in the herd. An area must be interesting if lots of scientists decide to go into it (not a view I share).

Another influence is the progress of simplifier science. Since questions about how the world works can be progressively answered, each plank resting on a previous one, then our list of great questions starts to shorten. We understand electricity now, and light, and geochemistry, and heredity. Not all and not perfectly, but we are not lost in the dark. This is a radical break from a century or two ago, and we will never regain that lost innocence (excepting their rediscovery if our civilization collapses). The remaining questions on our list become the ever harder ones, those we lack tools or language to attack, and have to burrow around in the roots of, hoping for local insights that illuminate the global puzzle. Consciousness, cosmology, and how the cell works are some of the leaders on our list, followed by others of more restricted domain. But there is no doubt the envelope of simplifier interests is shrinking, pursued by ever more scientists. Without being able to currently provide better numbers on the fraction of simplifiers and constructors, it seems to me that simplifiers are growing more slowly. For them, science becomes harder and more competitive, and the sense of decay in the distance looms.

This attitude was recently discussed in *The End of Science* by John Horgan, a writer for *Scientific American*, who surveyed elderly successful scientists in a range of fields who muttered about decay when pontificating on deep questions and their future. It caused a storm of protest from many scientists. The implication depends on missseeing all science as simplifier science, and as I have tried to show, this is far from the reality. Perhaps his truth is that the building blocks of science are not going to be crumbled to dust by a radical new approach or concept—our basis is now clear and solid in most spheres.

Constructors are not immune to their colleagues' anxieties, even if they don't separate themselves out in the explicit way I use here. Creative opportunities continue to grow as simplifier science adds more to the realm of the possible. Inventing new systems will never die. To see how little distance we have really come, imagine inventing plants with entirely new chemistries suited for other planets, which would provide a richness and diversity that could rival our own. Our imaginations remain so limited that nature is still far more interesting than we can conceive, and continues to provide endless surprises in riches. Science will not die from withering returns.

HOW DID WE GET HERE?

Ignoring the historical perspective, I have aimed in this book to capture a snapshot of how science is done in the present now. The problem in longer historical narratives is that they skirt key tangles that science is currently in, focusing on narrow slices with fewer actors and sagas, and often report from preprofessionalized eras. But the science ecosystem has indeed not always been the way it is now and has evolved (figure 9.3). Although none of the indicators such as the funding of science, the publication of papers, competition for the public ear, or the stability of institutions that I have looked at in this book shows any distinctive breaking point, I nevertheless emphasize they depict a state of some crisis that was not apparent twenty years ago. Why should it be that recent years have wound up the science ecosystem into an overtightened net?

I believe a key global factor in this is the inflation of scientific esteem, which has eroded as researcher numbers have expanded. The system of science is an extremely peculiar public good from a game theory point of view. Completely contrary to any economics perspective where secret information can be mined for profit, scientists give away their results, publishing them openly for anyone to access. In return for this service, society has promised them high esteem, and they have developed a pecking order built on this esteem.

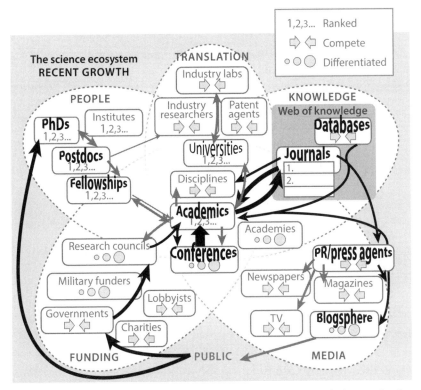

Figure 9.3: The most dominant changes within the science ecosystem over the last two decades. Darker text and arrows show main increases.

In the last decades, science esteem has been eroded in several ways. Continually rising individualism in societies (led by the United States, Australia, and the UK, but now rising in Asia as well) has focused people more closely toward financial rewards. In most countries scientists can make more (often much more) money outside research, but still many choose to remain inside the science ecosystem instead, preferring its intellectual stimulation. Increasing societal focus on financial rewards has translated for them into an increased focus on science esteem. At the same time, the growth in the number of scientists has led to an inflation of competition, diluting the pecking order of honor. This pressure acts together with the Internet, which has increased the aspirations of all scientists to higher esteem by showing them a much larger number of competitors.

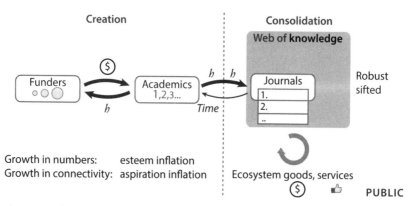

Figure 9.4: The division between the two parts of the science ecosystem, concerned with creation of new research, and its consolidation into the web of knowledge. The latter is less affected by intense pressures in the former, which are driven by inflation in the value of esteem by the increase in number and funding of scientists, as well as the growth in their connectivity.

Within the natural world, competition provides a point beyond which herds, packs, or prides break apart if they grow too large, or natural resources plummet, and a stable dominance hierarchy breaks down. By contrast, in science the division between performing science research, and accumulating the web of knowledge, so far limits how stresses are resolved. Inflation erodes the value of esteem, placing severe pressures on the former (doing) component of the ecosystem, but in the latter (knowing) sphere the results that are sifted through into the web of knowledge remain spectacularly robust (figure 9.4).

Over recent times, funders have provided more money and want more productive and higher-impact return from the researchers, while academics push to gain esteem (with h as one proxy metric) but have less time to read the burgeoning growth of results. The robust and sifted web of knowledge remains a key resource for society, delivering ecosystem goods (that can directly generate wealth) as well as ecosystem services of wider value for society. However I have shown how tightly tensioned the current system is, and how its increasing competition encourages hawk-like behaviors. We might be skeptical of star scientists' rosy views that more important science was done in the past, but it does seem that the interplay of ecosystem and culture influence the rate of science breakthroughs around the globe. The twinning of competition and globalization are not always beneficial to the development of science, as we start to fund too many of the same bandwagons, missing vital side discoveries, and focus research in the same ways favoring scientists of the same hawk-like mould. It remains unclear how much stress the science ecosystem can take, but this is an experiment that is indeed underway. In the final chapter I will ask what we might want to do about it.

CHANGING THE ECOSYSTEM

This was a chapter I was not planning to write. So far I have surveyed the science ecosystem as we find it, giving insights into what science you hear about, how you pay for it, and how scientists really do the work they do. My aim was to encapsulate in this book what I saw as the problems for science, not offer solutions whose controversy would cloud the mapping itself. However all who read my earlier drafts demanded suggestions. They insisted that given the discussions in previous chapters, I offer some vision of hopeful directions. The problem then for me is a stark choice between suggestions that are conceivably beneficial and those that are politically accessible; between pragmatism or vision. Before starting off though, let's think about some possible directions that might emerge if we persist in the current evolution of the science ecosystem.

In one vision, we can imagine that an internationally composed science board are tasked with creating a list of the research problems they want to be tackled. Their short-listed research bids would then be entered in a global reality TV contest, spinning messages to the public about what research will be done in order to gain points for public funding. This way the people who really fund science (all of us) could decide on what gets done. Slick videos, popularized science messages, celebratory presenters, visionary aims would all be pored over. Think Eurovision Song Contest meets BBC *Planet Earth*. It is plausible that on this global stage longer-term partnerships between universities and institutes in different countries would form, to combine skills, local resources, and science project leadership. Such conglomerates would make money selling newly minted technologies back to consortia of countries that invested in their new pet industries, and in these would focus the real science competition. In this future, I would be betting on my favorite research bids, at odds linked to current science fashions.

In another vision, such yearly lists of science problems might be disclosed to waiting teams to be auctioned off to the best cost-benefit proposal. Depending on the problems, different scientists would rapidly ally with each other, alliances advised by their local financial coordinators looking for low-cost partnerships. Projects would be identified by evaluator panels, guided by artificially intelligent crawlers across the web of science knowledge picking out gaps and conflicts, opportunities and puzzles. Their results would be attached onto this web of knowledge, visibly marking their contributions as detail, accretion, or underpinning. This would deliver science through problem-pull, rather than curiosity-push.

Conferences, now sponsored by journals, might combine presentations selected automatically based on the number of downloads of a paper, with electronic ranking of Twitter-submitted questions in real time (already trialed). High enough rated, and you get to annotate comments on the speaker's paper in real time, developing public debate with the speaker. All this develops in immersive virtual reality, with the corresponding ability to grab a local coffee at your iConference Center, while joining socializing groups within the virtual conference hub. Attending conferences would be now through competitive application, and you'll get thrown out if you don't strongly get involved, or if you check your e-mails instead of paying attention. Scientist will compete to attend the most prestigious events.

Perhaps in judging what science to fund, a new international body might be created, housed in a far-off location (a Pacific island), to which successful scientists are seconded for six-month periods. It's like the jury service in the United States and UK, and you can't get out of it if you've been funded before. Mingling with other scientists from different fields, removed from the hurly-burly of day-to-day research, and working together on developing areas of relevance gives a better perspective on science proposals. These science jurists are partnered with short-listed teams, to challenge and improve the proposed science. To encourage a long-term focus, jurists gain an index-linked pension stake from the science investment base on the world markets depending on the market-perceived value emerging from the programs they fund. Financial instruments based on these science investments also generate research funding. Through this tour of duty our jurists also reconnect with their initial passion for science.

Science careers might also be based through multinational conglomerates that act as science consultancies, hiring out postdoctoral researchers to funded projects. Giving an alternative career structure to the current academic route, talented researchers would command high salaries, and their managers would have incentives to add to their training in new areas. Combinations of skills in different fields would command a premium for the interdisciplinary expert, while some might specialize in crafting grant applications for scientists who have promising ideas but weaker project leadership.

These scenarios show other ways to conceive of a science system, and obvious extensions of our competitive, market-based science culture. Some seem to me dystopian, while others plausibly efficient. Mapping our science ecosystem helps widen our views of what can make alternative versions work better or worse, and guide our choices. This is not a minor aspiration because science has underpinned all our increased overall prosperity and improvement of the human condition for the last centuries—this collective investment is the future of *Homo Sapiens*.

WHAT TO CHANGE—A SMORGASBORD

How to change any system is a political question, but it is important to have some visions of where we might want to go. I list below eleven suggestions that I conclude from exploring the ecosystem of

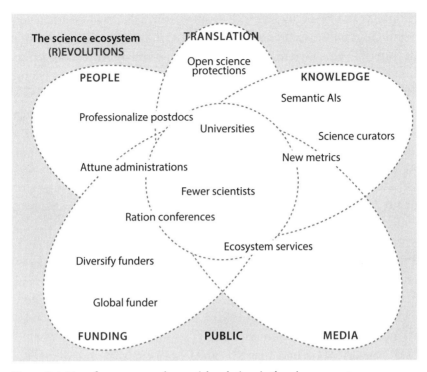

Figure 10.1: Map of some suggested potential evolutions in the science ecosystem.

science in writing this book, which can be graphically summarized across the science ecosystem map (figure 10.1). I divide these suggestions into those focused on competition, on connectivity, and on careers

HOW TO CO-OPT COMPETITION BETTER

Competition is increasingly used to ration the resources invested in science. In particular an alternative focus on people, metrics, conferences, and funding can be developed.

(1) Mitigate the effects of rising numbers of academic scientists globally.

As I have shown, funding ever more academics racks up the competition at every level and seems to produce ever decreasing direct science returns on investment. Less developed countries see benefits from workers with science skills, so this number will rise further, producing scientists indeed of value to their societies. Although funding for science will continue to rise in tandem (for reasons outlined in chapter 7), increased competition will not bring as much of an increase in results globally. Most scientists do not (yet) see this as a problem to tackle, but it threatens many aspects of the science ecosystem as outlined in earlier chapters. I hope it can usher in a golden age of science exploration, but it may just as easily erode the functioning of the ecosystem, dampening scholarship and privileging showmanship, while concentrating funding in elite groups and institutions, which is what a competitive model suggests. Science will not disappear, but its continuing health will be challenged.

No mechanism exists to even consider this issue. For young scientists it is a particularly emotive proposal, suggesting further barriers to their entering academic science careers. Perhaps partnerships between universities in different countries will enable a slow shift in researchers from one country to another, stabilizing the overall science population. With increasing success in science linked to bandwagons, science overcrowding can worsen quite rapidly so action is desirable. Can or should highly developed Western countries reduce their number of scientists, while retaining their investment in science? Can ethnic and gender diversity simultaneously improve within science, and can this help with many of the systematic issues such as communal inflexibility discussed throughout this book? Can we find ways to encourage scientific diversity, dampening bandwagons and fostering longer-term studies while retaining competitive peer review?

We might imagine that market forces could solve these problems (eventually), but as I have shown, the market for scientists within the ecosystem of science does not have traditional constraints. More robust discussion about how many physicists, biologists, chemists, and other disciplinary scientists any country needs must help.

(2) Develop competitive metrics to balance citations.

The competitive focus on citations drives particular styles of scientific progress and scientists who excel in this one dimension. Since metrics are the only current way to handle the information overload discussed in chapter 4, we must develop new metrics that emphasize other aspects needed in science: leadership, collaboration and cooperation, mentoring, communication, persistence, sharing, simplicity, and rigor. How researchers are seen in their communities must become more nuanced and collectable online, with combinations of "likes" now mixed with "respects," "acknowledges," and "recognizes." The rise of "network science" from consortia of scientists, which is already common in the fields of astronomy and particle physics, can be encouraged to spread further across science disciplines. Such consortia already demand measures beyond the h-factor recognition.

(3) Encourage anarchic ways to fund and support science.

The loss of diversity through globalization and the spread of "best practice" and bandwagons erodes variety in the types of science funded. The collection of units interacting inside the science ecosystem reinforce the current memome, providing an interlocking but stable system with little appetite for redirection.

New emerging players on the scene are philanthropists, and while many choose broad visions, such as sustainability, cosmology, searches for extraterrestrial intelligence, or eradicating malaria, among others, they are still influenced by bandwagon visions. We should encourage diversity not just in visions, but in the mechanisms used for selecting and developing science. Craig Venter, who aimed to patent the human genome (by trying to make money out of it from biomedical advances), stimulated new efforts as well as ruffled feathers. Anarchic schemes can shake up complacent traditions and are more likely to develop new types of science ideas.

Another collective of conformity to battle is the decreasing differentiation between species of institutions such as universities. Increased global competition between universities is leading to greater imitation—instead we should find ways to reward *different* ways of doing things. Growing numbers of partnerships between universities across the globe can purge diversity. Ecological theory notes that in order to coexist, species must have some maximum limit of similarity—they must be different from one another in some fundamental way, otherwise one species beats out the other and takes over the available niche. The danger that institutions of science will not retain fundamental differences is similar, that one type will take over and essential diversity will be lost.

(4) Encourage funders at a global scale.

An additional monolithic world funder of science is counter to many of the suggestions noted here, since it risks conformity. But it would free science from the notion that particular countries uniquely benefit from science, the misleading concept of competition between countries, and of the idea of science superpowers. The question of who funds and who decides what science is done is currently restricted to less than 10 percent of the planet. A global funder could form part of a system of science funding, driven partly by UN-sponsored societal goals.

HOW TO DEVELOP BETTER CONNECTIVITY

(5) Ration conference attendance.

The rise of science travel has few checks, and the system encourages ever more competition to present talks. Perhaps a serious development of virtual conferences will help, while some cap on the annual number for each researcher would encourage serious focus on choices. How can researchers be encouraged to value their own time more? Can we encourage conferences to focus more genuinely on their goals (even defining them would help!), which may suggest changing traditional session formats and bring in more outside influences to encourage broader exchanges. Useful disruptions to this niche are desirable.

One positive trend I have seen is for institutions and major charities to feel more at liberty to trial different ways of holding scientific meetings, particularly across interdisciplinary boundaries. For instance, the Royal Society of London (founded already in 1660 to broker argument) brings together select scientists who don't normally meet, in secluded convocations that chew over how their disciplines could focus together, such as neurobiologists and nanoscientists unpicking how the brain works. Such organizational experimentation is possible because failing has minimal consequences. Encouraging risk taking for developing and funding research directions or in publishing science are also highly desirable, but as we have seen the stakes are much higher for these aspects, and any steps need to win wide support in a contested environment.

(6) Develop artificial intelligences to mine the web of science knowledge.

Science is overloaded with information. The idea of a "semantic web" for science remains now trivially based on keywords and lists of publications. One key opportunity for the emerging generations of artificial knowledge systems is to tackle the wonderful horde of data and insights locked within scientific papers. It would enable much better connection of scientists to important problems, and enable negotiation through the vast information overload. This is a grand challenge problem that should be tackled cooperatively in the next decade and is completely achievable given the current advances in deep-learning methods. Despite a few nascent efforts, progress is neither prioritized nor rapid. It should be a priority for concerted transnational government funding.

(7) Encourage Open Science platforms.

These can provide a way to allow industries to better access science progress and science experts. We should investigate new ways to incentivize companies and academic collaborators to share science in public, by giving new forms of protections or revenue in return for delivering this public good. Experimental ideas in this space are already in progress, for instance with publishing forays into Open Access and Open Data (see chapter 4). Less advanced ideas are Open Instrumentation (sharing the technologies and know-how to build science equipment, often now possible using low-cost hardware), and Open Communities (trying to foster research tribes online, which has not really worked on a large scale from my experience). Optimistically one can see these increasingly bearing fruit. Governments have a role to encourage this since critical mass is essential, but philanthropic funding may give a vital spur.

(8) Identify and support science ecosystem services.

Ideas such as scientific beauty, elegance, playfulness, or intuition need a stronger recognition within the science ecosystem, despite their apparent vagueness. Can we formalize their quantification, actively measuring their health in components of the ecosystem? Seeing them as necessary parts of the system will help institutions, journals, conferences, governments, and companies find resources to devote to them. We should do more to identify and audit these, involving social scientists to help, despite the difficulties that are found in auditing any natural ecosystem services.

HOW TO DEVELOP BETTER CAREERS

(9) Develop consultancy companies for postdoctoral researchers.

The science ecosystem must tackle the continuing uncertain career structure for researchers, which is particularly damaging for gender and ethnic diversity. We should investigate building a new class of science consultancy companies that shelter and support postdoctoral researchers who are hired out to universities and research-intensive companies. This will provide a stabler development path, provide incentives for their further training, and better support their financial positions and rights. Another possibility is to foster this support within each university as a type of franchise. This is an open commercial opportunity that could be rapidly exploited.

(10) Make consistent frameworks for administering science.

In most organizations, researchers are tasked with many conflicting duties, and no overall prioritization given. Clear overarching goals and better clarity of management would allow organizations as a whole to understand how to enable science. An aim must also be to reduce the bureaucratization of research, which is increasing in all organizations, paradoxically because of their increasingly professional approach, which unfortunately is disconnected from research aims. Part of this would involve upgrading medieval accounting policies to treat grants more like investments.

(11) Fund science curators.

A new class of curator scientists should be supported who are linked to the cutting edges of fields, but whose job is instead to connect together different research teams, areas, technologies, and visions. Using these scientists to apply Occam's razor more consistently and wielding simplicity over exotic explanations ("techno-bullshit") would counteract trends to bandwagon science. They would require new electronic platforms, and new support mechanisms, as well as entirely new training pathways.

WHERE TO GO FROM HERE?

There are many questions I have found from my exploration of the science ecosystem that cannot be probed further, owing to lack of accessible data. I hope that I will have stimulated some of those involved around the ecosystem to feel compelled to seek answers (or help provide them at thesciencemonster.com). A more informed science of science fostering is needed.

I have throughout this book stressed the health of science in general and my fascination with all areas from simplifiers to constructors. However I have been more concerned with the way that science is encapsulated in a system that works at both an individual level and at the level of institutions. In some ways it may appear I have become cynical of the system that supported me so far in my career! More accurate would be that I worry about taking it for granted, in its not being able to support younger generations of productive and creative researchers intrigued to explore the world, who are crucial for the future health of our societies. This starts with fundamental misunderstandings about what typical scientists do, and how they are motivated. It also ignores the large range of structural forces that govern what they work on, how they work on it, and what we get to hear about that they have been working on. I believe the richer picture summarized here allows a more nuanced view of information we are all given about science.

Without competition science dies. But with competition, it skews. How do we go forward? The way we do science can surely change, even if we cannot see how or what will follow. The system of science sketched in this book has evolved to its current state but can jump into new shapes. It might happen dramatically through a massive loss of interest from young people that leads to government disillusionment to fund it, or it might evolve with the much vaunted and hoped-for knowledge economy liberating the masses from un-rewarding jobs (thus increasing competition for careers such as science). Scientists and public alike have lacked a bird's-eye view of how the system works, which has been my motivation here. Only from this point can we ask how we might change it. The science ecosystem is now in your hands too.

INDEX

Page numbers in italics refer to illustrations in the text.